D0758734

Energy Storage

Energy Storage

J. Jensen

NEWNES-BUTTERWORTHS

LONDON · BOSTON
Sydney · Wellington · Durban · Toronto

The Butterworth Group

United Kingdom Butterworth & Co (Publishers) Ltd
London: 88 Kingsway, WC2B 6AB

Australia Butterworths Pty Ltd
Sydney: 586 Pacific Highway, Chatswood, NSW 2067
Also at Melbourne, Brisbane, Adelaide and Perth

Canada Butterworth & Co (Canada) Ltd
Toronto: 2265 Midland Avenue, Scarborough,
Ontario M1P 4S1

New Zealand Butterworths of New Zealand Ltd
Wellington: T & W Young Building,
77–85 Customhouse Quay, 1, CPO Box 472

South Africa Butterworth & Co (South Africa) (Pty) Ltd
Durban: 152–154 Gale Street

USA Butterworth (Publishers) Inc
Boston: 10 Tower Office Park, Woburn, Mass. 01801

First published 1980

British Library Cataloguing in Publication Data

Jensen, Johannes
Energy storage.
1. Energy storage
I. Title
621.4 TJ165

ISBN 0-408-00390-1

Typeset by Butterworths Litho Preparation Department
Printed in Scotland by Thomson Litho Ltd.,
East Kilbride

Foreword

By Professor Michael E. McCormick, Consultant to the US Department of Energy

Technology in the twentieth century has advanced dramatically due, in no small part, to the efficient utilisation of oil as a primary fuel. Within the past several years we have become aware of the fact that oil is a depleting resource and, by the mid-twenty-first century, this resource will no longer exist in significant quantities. Thus, in the present decade we have started collaborative efforts to seek methods of conversion, storage and utilisation of alternative energies. While much attention has been devoted to the development of energy conversion techniques, energy storage has somewhat been taken for granted, although significant advances have taken place in the energy storage field.

The technical community has been waiting for a publication which both introduced the various energy storage techniques and outlines their utility. *Energy Storage* is thus a welcome addition to the technical literature. The topics covered herein are presented in a concise and to-the-point manner which enables the reader to easily understand the applicability of each energy storage technique.

The author, Dr Johannes Jensen, is an internationally known scholar/scientist/engineer who is eminently qualified to write on the subject. He has performed an excellent service to the technical community in writing this book.

Preface

The development of energy storage systems has become increasingly important as the shortage of petroleum fuels and the intense concern over air pollution makes effective use of available energy essential.

The installation of energy storage units on electric utility networks will permit the utilities to store energy generated at night by coal-burning or nuclear baseload plants and release the stored energy to the network during the day when demand is highest. This reduces the need for gas or oil burning turbine generators.

Large scale utilisation of the renewable energy sources such as the sun and the wind depends on storage facilities, since these sources are variable over the day and over the year.

In the transport sector the demand for new ways to store energy has become evident. In a few decades the filled petrol tank in our cars will have to be replaced by other storage units if we are to escape from the present dependence on oil as a primary fuel.

This explains in a few examples why a book on energy storage is needed. The book sets out to present a broad review of the very complex subject of energy storage and to enable the reader to choose the proper storage method for a particular situation. The choice of storage method depends on many factors, some of the most essential being:

the time for which storage is required
the quantity of energy to be stored
the form of energy needed for consumption
variations in the rate of consumption.

Also, this book describes basic theory and calculations of key parameters such as energy density and power density, and covers such fields as:

heat storage
chemical storage
mechanical storage
electrical and magnetic storage.

A chapter on existing storage systems as well as useful tables of the most important energy storage parameters is also included.

Energy Storage is intended for engineers, applied scientists and students, though not necessarily specialists in the field of energy storage. It addresses industry, science and research, and it is hoped it will be useful not only to professional people working on energy conservation, utility load levelling, and stand-by or emergency power problems etc., but also to the individual concerned about a storage unit for use with the solar panel on his home. An attempt has been made to make the book useful in engineering offices by including a list of manufacturers of energy storage equipment.

The author has had very valuable discussions with a large number of individuals in Europe, USA, and Japan, in particular with colleagues within the EEC energy conservation programme. The time and ideas generously contributed by all these individuals are much appreciated. Acknowledgement is also made to the many companies and research institutions who provided data and illustrations.

J. J.

Contents

1 Introduction

Energy storage is, in one way or another, part of all events both in nature and in man-made processes. There are many different kinds of energy storage systems, some containing large amounts of energy, and others very little. Some are part of energy transfer processes and others are part of information transfer systems. Such a variety of possible applications clearly means that several key parameters must be considered and that they differ from one application to another.

A classification of energy storage systems therefore always tends to be very complex. In most cases two features of the systems are crucial:

the amount of energy to be stored
the length of time for maintaining the storage.

Energy density and storage time are the key parameters to be considered when discussing storage systems.

In nature we find examples of storage systems containing both large and small amounts of energy. Storage times also vary greatly. Take, for instance, the information transfer system in animals, including human beings. In this system, called the nervous system, the transfer of impulses or signals is accomplished with minimum energy consumption, and since information has to be transferred rapidly, the storage time at different locations along the impulse path is also very short. The liver, on the other hand, contains a relatively large amount of energy, and it can act as an energy store for the body over a long period of time.

A similar variety is found in man-made systems. In a high frequency oscillator, where a small amount of energy is transferred periodically from a magnetic field to an electric field and back again, the storage time in each field is a fraction of a microsecond. In insulated hot water storage tanks for solar heating panels, on the other hand, the energy content is high and the storage time has to be several months. The energy has to be stored from the summer until winter when the demand for heating appears. A battery for a pacemaker with power output only in the milliwatt region has to last 5–10 years, and a peak shaving storage unit for electric utilities must be able to reach power outputs in the hundreds of megawatt range, but only for a few hours.

In this book an attempt is made to describe and assess mainly large scale storage, i.e. storage in energy transfer systems. However, some of the principles of information storage will be discussed, not only because of scientific interest, but also because some of the components now being used in information systems may well in the future become part of energy systems, due to advances in technology and availability of new materials. The storage problem may involve energy which is used directly in the same form as it is initially available, or may involve conversion into other forms. In both cases, the problem of the required time of storage must be considered.

Historically, the energy storage problem was solved by piling lumps of wood together, or by damming streams to provide a working head for a waterwheel (*Figure 1.1*). Later, the more concentrated form of fuel, coal, became the most important energy store. Nowadays we are accustomed to the filled oil tank as a most convenient form of energy storage. Oil-based fuels offer ease of use, availability and relatively low price. Storage of oil is easy and the storage time simply depends on when the tap of the tank is turned on. Storage can be maintained without any losses and the energy density is high. Oil can be used as a source both for power and heat and also for transportation as well as stationary applications.

However, oil resources are limited and despite the fact that the exact time when oil resources will run out is at present a matter of dispute, everyone agrees that a substitute for oil products as energy stores must be developed in the near future – and here we are talking about a number of decades only.

The oil supply will in a short time reach a maximum and oil consumption will then go back to the normal historical level, namely zero. This, of course, is an unusual view of the energy situation, but if an historic time scale is considered one must admit that apart from some minor use of crude oil for road and building

Figure 1.1
Once wood piles and watermill dams were the only practical means of long term energy storage

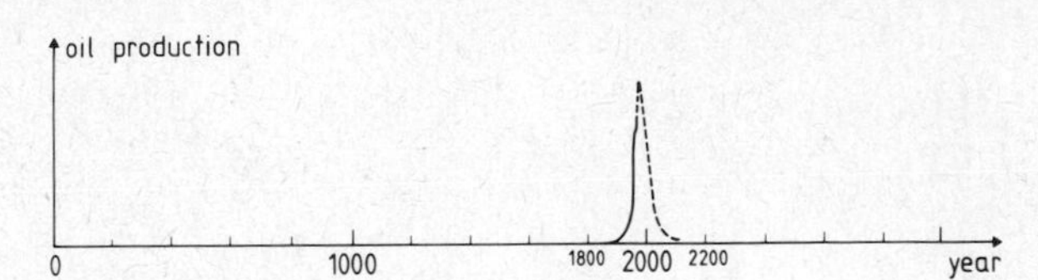

Figure 1.2
Historically the normal oil production is zero

applications in ancient Babylonia, the oil adventure began when oil was struck in August 1859, at which time the first drilled oil well was utilised at Titusville, Pennsylvania, USA. The production and consumption of oil has since increased rapidly but on an extended time scale oil usage will appear as a sharp peak.

The future decrease in oil production will depend on many factors, but it is unquestionable that in a small number of decades the easy-to-utilise oil resources will have been used, even though remaining resources may be reserved for transportation and for the chemical industry. Oil shortage must not be considered as an unusual and special situation. Rather, availability of oil is a special condition for the 20th and the 21st centuries, and although known oil wells may contain larger quantities than now estimated and the cost of exploiting wells at exotic places will probably decrease, there is an end to the oil adventure. Therefore, and especially because oil has been a readily available primary energy source and

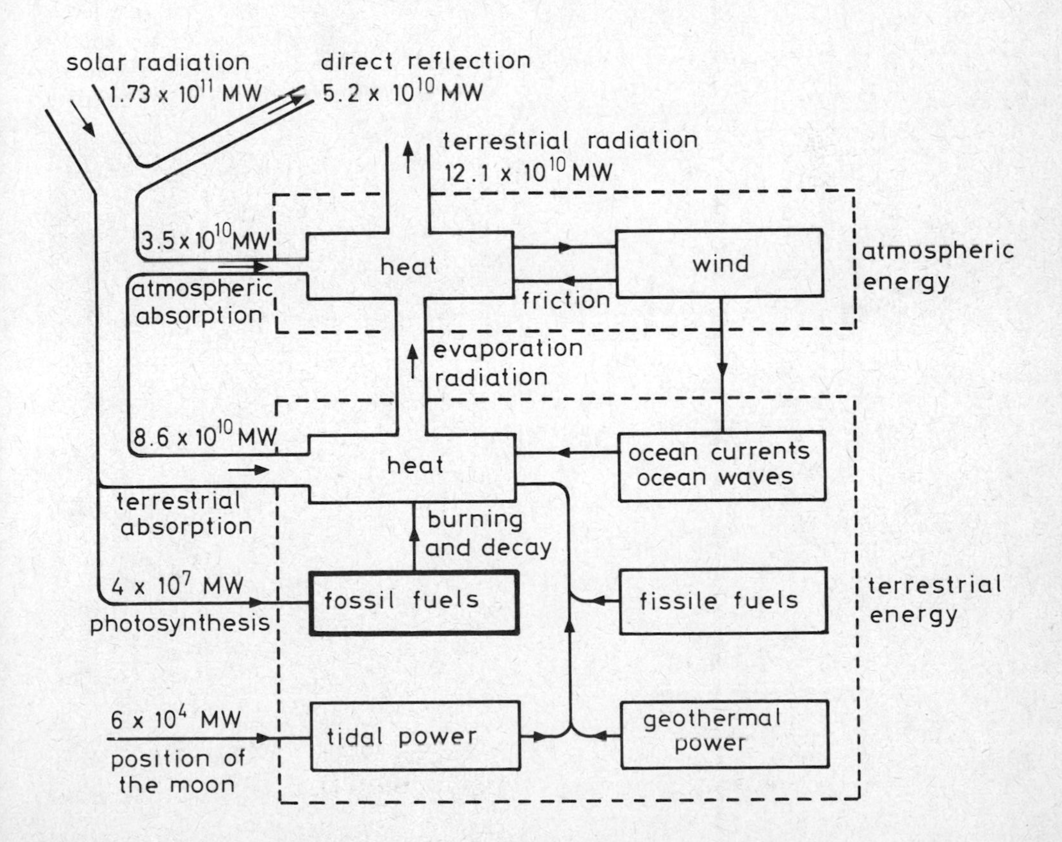

Figure 1.3
Global energy flow and storage

a good way to store energy, international research efforts in recent years have been directed towards the search for alternatives.

One obvious way to attack the problem of finding new primary sources and new kinds of energy storage is to look at the global energy picture shown in *Figure 1.3*.

The picture would not be complete without the fusion fuels, and a sub-storage box for hydropower in the evaporation flow ought also to be included for completeness. The essential point however is that the boldly outlined box 'fossil fuels', i.e. crude oil, natural gas and coal, represents both the main primary energy source and the best forms of storable energy.

Attempts to utilise some of the other energy flows – and storages – are well known. Recently the renewable energy sources have attracted considerable interest. However, most renewable sources do not provide a constant energy supply and, unlike fuels, can not be directly stored; they require secondary storage systems. There is another important difference from the fossil fuels, namely energy quality.

2 Energy Quality

High quality and low quality energy

The theory of thermodynamics tells us that the quantity 'entropy' is used to describe energy quality. Entropy is a rather difficult concept but, fortunately, it is possible to explain in another way what is meant by high quality energy and low quality energy.

Most of us have a clear feeling of the distinction between electrical power and heat; surely we would not be in any doubt of the difference between the failure in supply of one or the other energy form. Failure of the heat supply could, for a short period at least, be handled by wearing a warm knitted sweater, but if the power supply — the electricity supply — fails, then industry and society as a whole ceases to function.

It seems obvious that our cars cannot run on a hot water bottle with tepid water as energy storage. Lukewarm water is low quality energy and in the transportation sector high quality energy is a necessity. By high quality energy we mean electrical energy, mechanical energy and some forms of chemical stored energy, e.g. the fossil fuels. By low quality energy we mean low temperature heat.

There is a rule, although a little simplified, which states that storing of heat saves energy while storing of electricity saves capital investment. Since the energy sector on a global basis faces a resource crisis both in respect of raw energy and capital, there are good reasons for investigating storage of low quality as well as high quality energy.

High quality energy and low quality energy are measured in the same units, and one does not distinguish the two forms of energy by the amount of energy itself. The amount of energy can be described in several systems of units (see *Table A.3*). Two of the most commonly used units are the Joule (J) and the kilowatt hour (kWh).

One kilowatt hour

Most forms of energy are sold without any information about the essential ingredient, the energy content. We buy it as barrels of oil or tons of coal or gallons of petrol, but this does not tell us how many kWh we are buying. Therefore let us consider what 1 kWh is.

It is the equivalent energy:

to 86 g of oil
to what we pay 3 p for if we buy it as electricity
to what we pay 1.5 p for if we buy it as oil

required to lift 1225 kg 300 m (a small car with 4 passengers up to the top of the Eiffel Tower)
required to heat 10 litres of water from 14 to 100 °C.

It is also equivalent to the mechanical work performed by an average adult person in half a day. So if we should return to slavery, to get all the work done by hand in western Europe, Japan or USA, we would have to have more than one hundred slaves each, hence that possibility does not exist – luckily enough, we may add.

The two jobs of lifting a small car up to the top of the Eiffel Tower and of heating 10 litres of water to 100 °C require the same amount of energy, but the former is certainly the more difficult. It requires high quality energy, whereas heating 10 litres of water to boiling point can be done by means of low quality energy. High quality energy is, so to speak, more worthy or more valuable and more usable than low quality energy.

High quality energy, for example electricity, can easily be used to produce low quality energy such as low temperature heat, whereas the opposite is much more difficult and in some cases impossible in practice. The flexibility of energy systems is therefore highest when storage possibilities for high quality energy are available, but as we shall see in the next chapter, there is a demand both for electricity storage and for heat storage. A solar panel system must have a storage unit in order to function, and the development of combined heat and power with heavy interlockings between the two energy qualities imposes a demand for both forms of energy storage.

Energy conversion

We have seen that high quality energy is more valuable than low quality energy, and therefore conservation of high quality energy in transfer processes is important. This certainly is true when transferring energy in and out of a storage unit. Such a process always involves some degree of energy conversion and usually the related fraction of low quality energy is called process-loss or waste. We call it waste because we do not use it, and not because that part of the total energy has disappeared. The first law of thermodynamics – the law of energy conservation – tells us that energy cannot disappear.

So energy conversion never involves a change of total energy and we have to look for another quality and that is 'entropy'. Energy conversion in practice always causes a change in entropy and we can divide the many different ways of converting energy into three groups:

1. Conversion of high quality energy into low quality energy, e.g. oil or gas burners for heating systems
2. Conversion of high quality energy into mechanical energy and low quality energy, e.g. combustion engines or turbines
3. Direct energy conversion.

Process 1 is, as we have discussed before, a conversion of high value energy into low value energy. Process 2 – the heat engine – is a conversion of high quality energy into useful mechanical work, and the efficiency with which this is done is determined by the so-called Carnot limitation. As an example we can calculate the theoretical efficiency of a turbine of a power station, considered an ideal four-step Carnot cycle, and the entropy change involved when the inlet steam temperature is, say, 600 K and the outlet temperature – the temperature of the condenser – is 300 K.

Entropy is defined as

$$\Delta S = \int_{T_1}^{T_2} \frac{Q}{T} \, dT \tag{2.1}$$

The ideal Carnot cycle is shown in the isotherm-adiabat plot in *Figure 2.1*. In practice such a process is impossible since it would

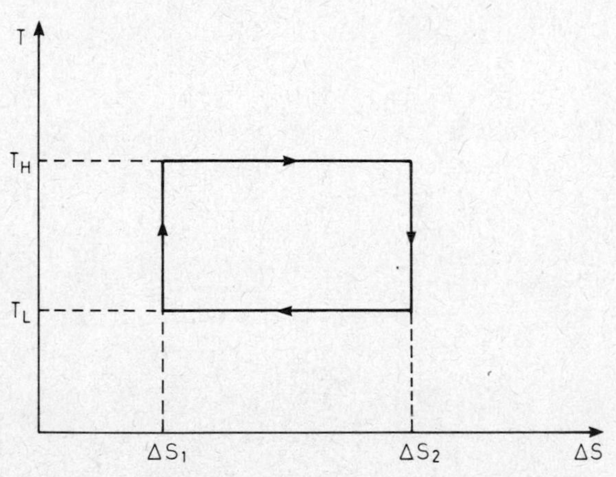

Figure 2.1
Isotherm-adiabat plot of the ideal Carnot cycle

require alternating an infinitely large heat reservoir and a highly insulated system, and this means that the efficiency in practice is lower than the theoretical maximum efficiency η_m, where η_m is defined as the useful work we get from the system divided by the total amount of work supplied to the system.

$$\Delta W_S = \int_{\Delta S_1}^{\Delta S_2} T_H d(\delta S) = T_H (\Delta S_2 - \Delta S_1) \tag{2.2}$$

$$\Delta W_R = \int_{\Delta S_1}^{\Delta S_2} T_L d(\delta S) = T_L (\Delta S_2 - \Delta S_1) \tag{2.3}$$

$$\Delta W_U = \Delta W_S - \Delta W_R = (T_H - T_L)(\Delta S_2 - \Delta S_1) \tag{2.4}$$

$$\eta_m = \frac{\Delta W_U}{\Delta W_S} = \frac{(T_H - T_L)(\Delta S_2 - \Delta S_1)}{T_H(\Delta S_2 - \Delta S_1)} = \frac{T_H - T_L}{T_H} \qquad (2.5)$$

where

T_H = high temperature
T_L = low temperature
W_S = total work supplied to the system
W_R = work removed from the system
W_U = useful work.

If we put the values from our example of the power station turbine into the formula for η_m we get

$$\eta_m = \frac{T_H - T_L}{T_H} = \frac{600 - 300}{600} = 0.5$$

With these temperatures the thermodynamic loss is 50 % or, in other words, the maximum efficiency is 50%. In practical systems, the efficiency of the conversion of heat into mechanical energy is, as mentioned, lower.

Note that the symbol W is used for work and energy. Energy and work are measured in the same units, joules (J), and in each case the index to W indicates whether it is input or output work or energy. We use W in order to distinguish it from E, which is used as the symbol for the electric field.

Work δW is the product of a quantity δq moved across a potential difference ΔP. Some expressions of this kind will surely be recalled from high school physics. Electrical work δW, when an electric charge δq is moved across a voltage or potential difference ΔV, equals the product $\Delta V \delta q$; the mechanical work when a mass δm is moved over a gravitational potential difference Δgh (h is the height and g is gravity) equals the product $\Delta gh\,\delta m$. In general we can write

$$\delta W = \Delta P \delta q \qquad (2.6)$$

where

ΔP = potential difference
δq = quantity.

In thermodynamic processes, e.g. heat engines, the quantity which is transferred is entropy and the potential difference is the temperature difference.

$$\delta W = \Delta T \delta S \qquad (2.7)$$

We have already used Equation (2.7) when deriving η_m and we will now in *Figures 2.2* and *2.3* look a little closer at conversion processes 1 and 2 mentioned on page 6.

Figure 2.2 illustrates the irreversible process 1 where the high flame temperature (1000–1300 °C) of a domestic gas or oil burner is 'converted' to heat the water in the radiator system to around 70–80 °C. The amount of heat ΔQ_1 at the flame temperature

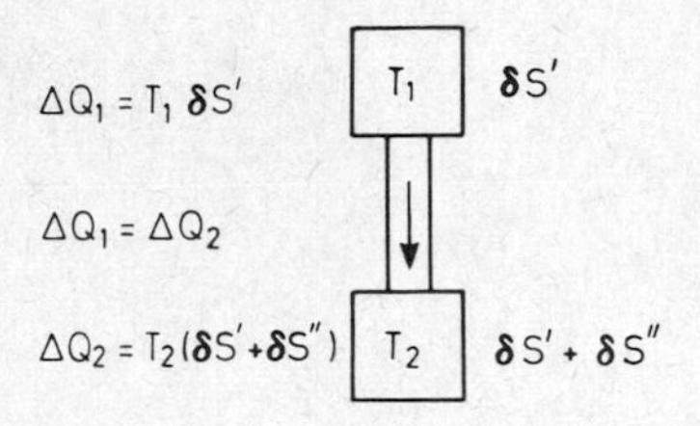

Figure 2.2 Conversion of high quality heat at temperature T_1 into low quality heat at temperature T_2

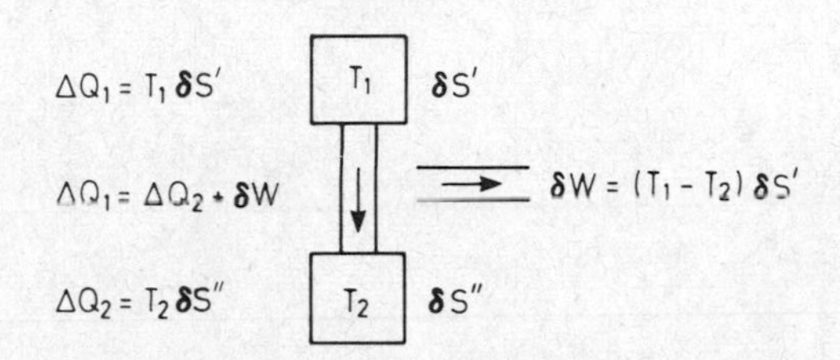

Figure 2.3 Conversion of high quality heat into useful work and low quality heat

is exactly the same as ΔQ_2 at the hot water temperature T_2. The only result is an increase in entropy from $\delta S'$ to $\delta S' + \delta S''$ and that is really of no use whatsoever. In terms of energy quality it is foolish to convert a storage of oil into a storage of hot water.

Figure 2.3 shows what happens in process 2. Here we do get useful work out of the initial high quality heat, and it is clear that process 2 is to be preferred to process 1.

Modern attempts to increase the amount of high quality energy, which results from converting chemical fuels, have centred on the idea of eliminating the heat stage entirely, under the slogan 'avoiding the heat trap'. Conversion processes without the Carnot limitation have been investigated. The fuel cell (see Chapter 5) is an example of avoiding the heat stage.

In theory it might be possible to attain thermal efficiencies close to one by the use of high inlet temperatures T_H. This would, however, require a temperature of several thousand degrees and in practical heat engines is not attainable because of the materials problems involved.

Fortunately, there are many ways of getting high efficiencies for the transfer of energy into and out of a storage unit. Pumped hydrostorage and flywheel systems for example exhibit overall efficiencies between 0.7 and 0.9. In flywheel systems the energy is kept in the same form (kinetic energy) and friction losses are minimised. It is important, when dealing with high quality energy, to keep as much energy as high quality energy during and after the storage process.

3 The Demand for Energy Storage

The effective use of existing, depletable fossil fuels and the utilisation of renewable alternative energy sources demands economical and efficient energy storage with flexibility in operation and siting. Availability of suitable energy storage systems will have a favourable impact on areas such as:

1. Utility load levelling – to improve load factors, reduce pollution in populated urban areas and to make better use of available plants and fuels
2. Storage for combined heat and power systems – to improve overall efficiency by offering optimum division between heat and power irrespective of load demands
3. Storage for electric vehicles – to replace petrol in the long term, reduce urban air pollution and improve utility plant factors
4. Utilisation of solar energy in its various manifestations – to relieve the burden on the finite fossil fuel resources and to improve the living environment
5. Storage for uninterruptable power supplies – to improve the reliability of supply for critical applications such as hospitals and computing facilities
6. Storage for remote location facilities such as telecommunication and meteorological stations
7. Storage for industrial mobile power units – to provide better working conditions, especially in confined areas such as warehouses, mines, etc.

This broad range of possible applications for energy storage devices is unlikely to be satisfied by a single method. In addressing the demand for development of storage devices and systems with different technical and operational characteristics, two quite different groups of applications are considered. These are stationary applications and transportable applications.

Stationary applications

Electric power systems have a well established need for large-scale energy storage which is being met in part by pumped hydroelectric storage. The full potential of utility energy storage can be attained only through development and implementation of more broadly applicable storage technologies.

The variation of load throughout the day, week, and year imposes a demand for storage especially with the increase in the use of large coal or nuclear plants designed to operate at maximum efficiency on base load and a future increase in utilisation of variable energy sources such as solar, wind, ocean energy etc.

Figure 3.1 shows a typical weekly load curve of a utility with and without energy storage. As illustrated by the upper curve, intermediate and peaking power involves extensive generating capacity. The load variation shown here is typical of the US situation, but it applies to most other countries, where cheap off-peak electricity rates exist. In countries where this is not the case the daily variation tends to be larger. In any case it appears to be the fact all over the world that installed capacity is about double the yearly average load.

If large-scale energy storage were available as illustrated by the lower curve of *Figure 3.1* then the relatively efficient and economical base load generation could be increased and the excess beyond off-peak demand (lower shaded areas) could be used to charge the storage system.

Discharge of the stored energy (upper shaded areas) during periods of peak power demand would then reduce or replace fuel-burning peaking plant capacity, thus conserving fuel resources. In addition the higher base-load level would replace part of the intermediate generation. Assuming that new base load plants use

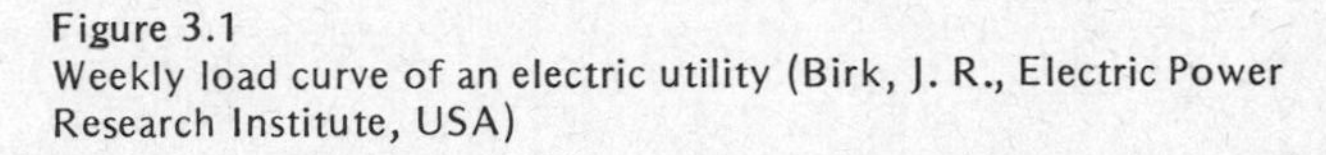

Figure 3.1
Weekly load curve of an electric utility (Birk, J. R., Electric Power Research Institute, USA)

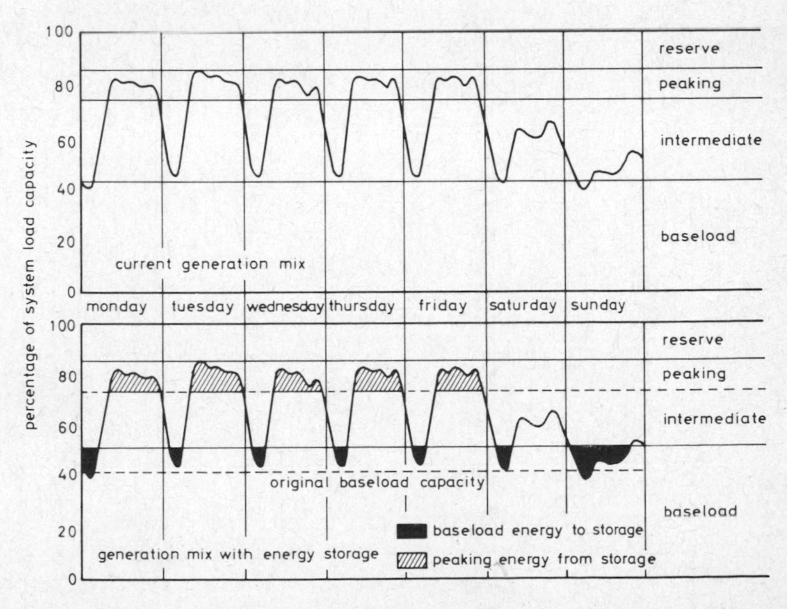

non oil-based fuel, there are further savings of both cost and of oil resources.

Use of energy storage to generate peaking power in this manner is termed 'peak shaving'. Load levelling describes the more extensive use of storage to eliminate most or all conventional intermediate cycling equipment. Energy storage efficiencies in the example shown in *Figure 3.1* were estimated at 75%, and if so the result may include an overall energy saving but not necessarily. The outcome depends on whether the higher efficiency of base load plants compared with peaking and intermediate equipment makes up for the storage inefficiency. In any case, a reduction in installed capacity is achieved.

The demand for storage in the case of combined heat and power production arises in periods throughout the day where relative demand for heat and power output cannot be met by the utility plant. At night the demand for electricity is usually very low, but the demand for domestic heating is high. A heat storage unit, e.g. a highly insulated hot water tank with few hours storage time capacity in this case, is essential.

Another major demand for stationary storage arises from the utilisation of renewable energy sources. These sources, which directly or indirectly relate to the solar radiation arriving on the surface of the earth, all vary with time. The variation of solar radiation itself is shown in *Figure 3.2*, from which the demand for long term heat storage is obvious. The demand for heating occurs in periods with a lack of solar energy, hence the storage time must be several months for solar panel heat storage. Also solar cell electricity requires storage especially because the present high cost of solar cells makes an optimum system design including a large battery attractive. The variation is less for indirect solar sources such as wind and ocean waves, and they fit much better to load demands throughout the year. Wave energy, which is essentially accumulated wind energy, fits the load demand quite

Figure 3.2
Energy of solar radiation on a clear day and on horizontal plane for different latitudes (Duffie, I. and Beckman, W., *Solar Energy Thermal Processes*, Wiley, 1974)

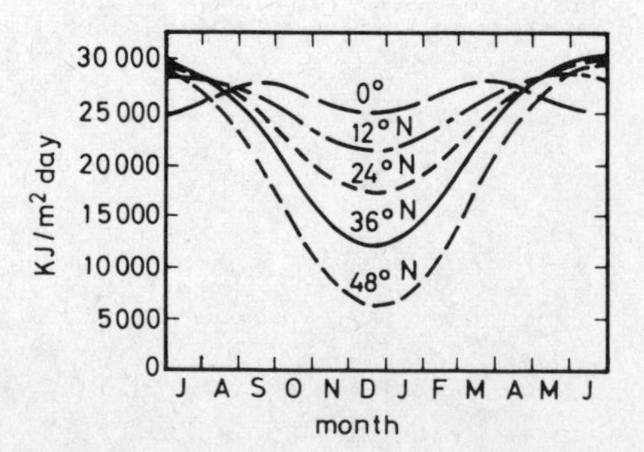

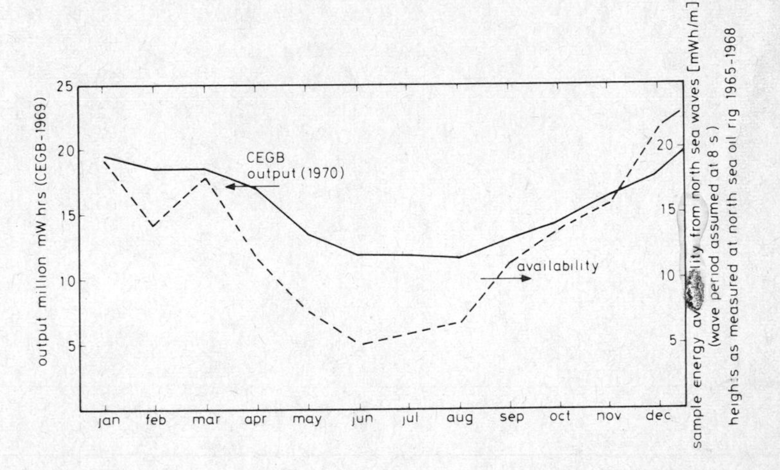

Figure 3.3
Comparison of monthly variation of power generation in UK and availability of wave power from North Sea waves (East Kilbride National Engineering Laboratory, UK)

well (see *Figure 3.3*) and wave energy installations might be connected to the utility grid without use of storage units. Both wind and wave installations impose a demand for storage in remote locations, where grid connections are not economical.

Transport applications

As already mentioned, one of the clearest examples of the demand for transportable stored energy is the need for an alternative to the filled petrol tanks in our cars. Whilst the main constraint for most stationary applications is low cost, there are, for the transportable applications, some additional technical constraints such as power density and energy density, both with respect to volume and weight of the storage unit.

In general, a good storage system for transport applications must meet, at least, the requirement referred to in *Figure 3.4* and it must also be reasonably safe to handle and to operate. Most of the effort during recent years to bring about an alternative to petrol driven combustion-engine-vehicles has been concerned with the electric battery vehicle, for which the major demand is a battery with better energy density than that of the lead acid battery (see Chapter 5).

The development of the transport sector has enlarged the demand for alternative fuel and storage forms. *Figure 3.5* shows some alternatives and it also shows that trends in the use of primary fuels and storage are in opposite directions to those of transport forms. The decrease in rail transport, where non oil-based electricity via overhead wires could act as the energy source, leaves an increasing part of the transport work to be done by vehicles with

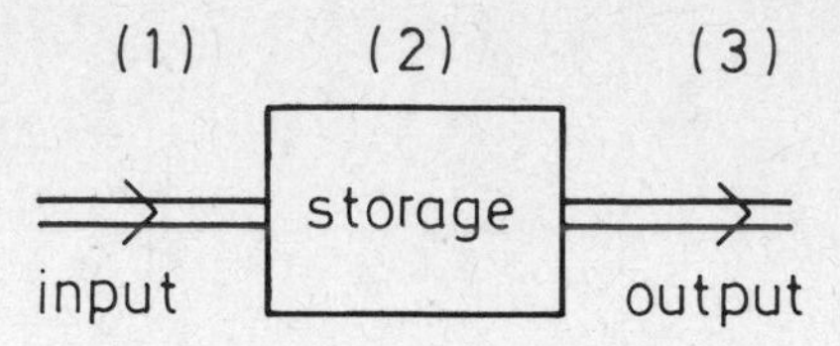

Figure 3.4
Requirements for a good storage system for transport applications. (1) High rate of energy flow with high efficiency for the input. (2) High energy content for sufficiently long time. (3) High rate of energy flow with high efficiency for the output

energy storage, which means an increasing demand for synthetic liquid fuels for heavy duty transport and for electric batteries for urban transport.

The range of a vehicle is determined by the energy density of the storage unit and the acceleration by the power density. Braking energy recuperation also depends on availability of a storage unit with high power density. The demand for very high power (5–10 times that needed to maintain average vehicle speed) is only required for short periods of acceleration and braking, and the ideal power supply for transport jobs looks very much like the one we find in nature, such as illustrated in *Figure 3.6*. For a

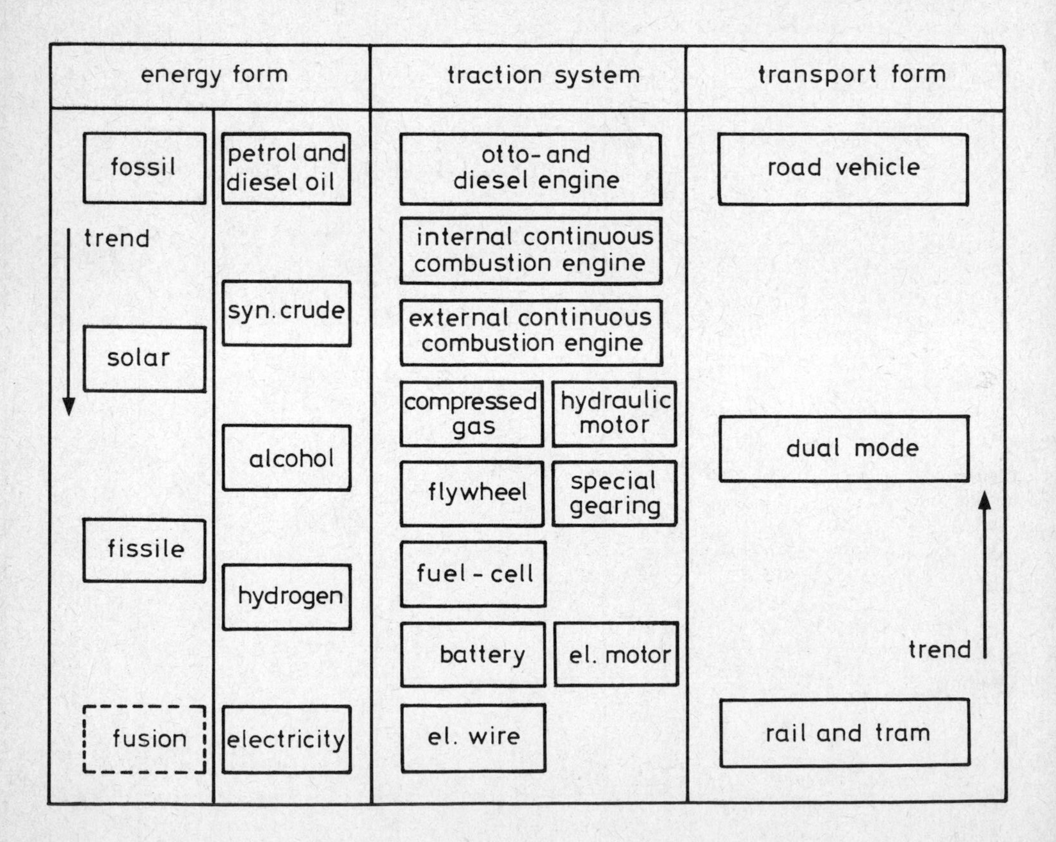

Figure 3.5
Trends in energy form and in transport form (Helling, J. and Jensen, J.)

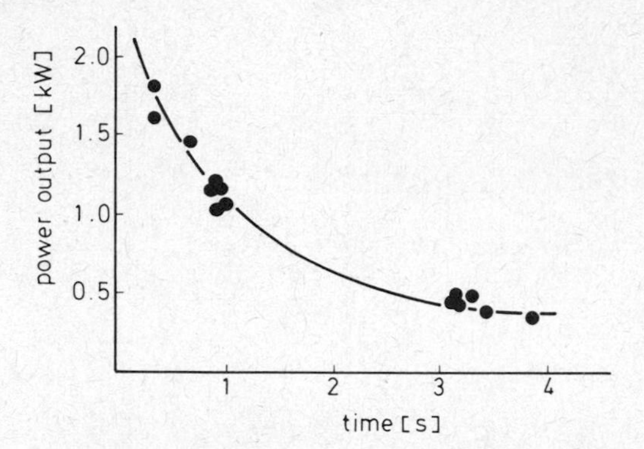

Figure 3.6
Extremes of power output developed by athletes during short bursts of activity (Edholm, O. G., 'The Biology of Work', Weidenfeld and Nicolson)

fraction of a second an athlete is able to provide about five times as much power as he can maintain for several seconds based on about the same 'energy density'.

The high power density requirement and the high energy density requirement is hard to meet with just one storage unit. The reason why current petrol systems meet both requirements is that the combustion engine is heavily oversized compared to the average power needed. Consequently, in order to design a practical power source, interest during recent years has been focussed on systems with more than one storage and conversion unit.

Integrated energy systems

In order to meet the previously mentioned requirements for storage and conversion units, it is necessary to do some energy system analysis, as no single unit will meet all the requirements. Two quite distinct systems are the hybrid system and the combined system. We shall define a hybrid energy system as a system with one kind of energy output and two or more energy sources as input. The combined system is the opposite, namely a system having one primary source and two or more different kinds of energy output. *Figure 3.7* and *Figure 3.8* show the two systems, S being input sources and C energy conversion.

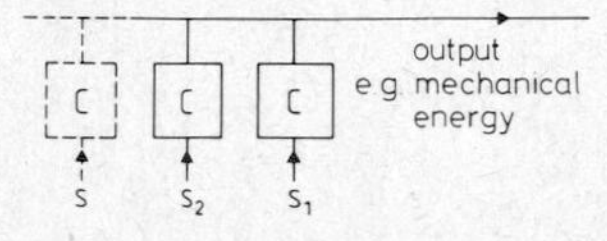

Figure 3.7
Hybrid energy system

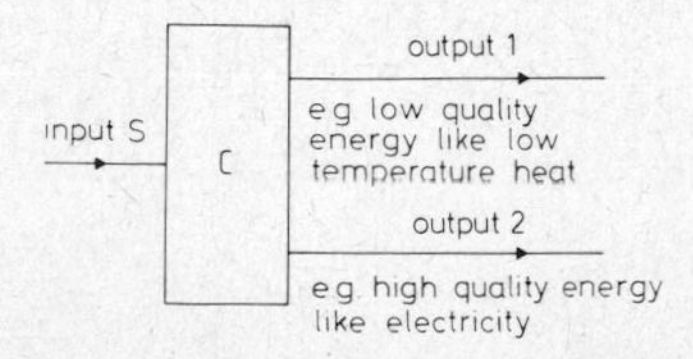

Figure 3.8
Combined energy system

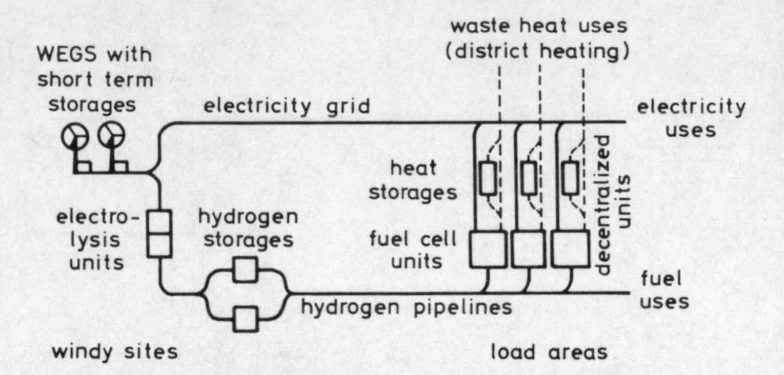

Figure 3.9
Combined system as energy storage for wind energy (B. Sørensen, Niels Bohr Institute, Copenhagen)

The hybrid system has been used in the transport sector and actual road tests have shown reduction in fuel consumption up to 40% for hybrid vehicles following an urban driving cycle. Since there are at least two storage units, one can be designed for high power density (acceleration and braking energy recuperation) and the other for high energy density (range).

The combined system has also been widely used in combined heat and power generation, where power stations utilise the waste heat from electricity production as district heating. Another example of a combined system is shown in *Figure 3.9*. The input source is electricity from a wind energy generator (WEG) and the system combines short term and long term storage.

Integrated systems using storage methods in suitable combinations make it possible to meet the energy demand from the available supply with high overall energy efficiency.

Further reading

Duffie, I. and Beckman, W., *Solar Energy Thermal Processes*, Wiley, New York (1974)

Helling, J., *Unkonventionelle Transport Systeme*, Institut für Kraftfahrwesen, Aachen (1976)

Ramakumar, R. 'Survey of Energy Storage Techniques', IEEE Region Six Conference (1976)

Sørensen, B., *Renewable Energy*, Academic Press, London (1979)

4 Heat Storage

Direct storage of heat in insulated solids or fluids is possible at low temperatures, but energy can only be recovered effectively as heat. Conversion to other forms, such as mechanical or electrical energy, will be very inefficient because of the fundamental thermodynamic limitations. Thermal energy storage, however, is ideally suited for applications such as space heating, where thermal (low quality) energy is required.

Two distinct storage mechanisms are

1. sensible heat storage, based on the heat capacity of the storage medium
2. latent heat storage, based on the energy associated with a change of phase for the storage medium (melting, evaporation or structural change).

Storage media proposed include water, rocks, and several salt hydrates. Hot water is, because of low cost, the preferred storage medium when the heat energy is to be used at temperatures below 100 °C. Other materials such as gravel or soil have been considered for use with heating systems.

For applications in other temperature ranges, materials such as scrap iron may be preferred where, as in rock-based materials, the temperature would not be limited by the boiling point of water. For the high temperature applications, the transfer fluid usually is hot air blown through the porous material.

The main problems of store design are

1. to establish a suitable heat transfer surface in order to get a fast transfer of heat to and from the storage unit
2. to avoid heat loss flow to the surroundings in order to obtain a leakage time which is large compared to the required storage time.

The heat loss from a storage tank depends on the tank surface area, and the overall storage capacity depends on the volume of the tank. The surface area is proportional to the second power of the tank dimension and since the volume is proportional to the third power, large storages need proportionally less insulation than small ones. The stationary temperature distribution $T/T_k = f(x)$, x being the distance from the centre of a spherical storage volume kept at constant temperature T_k, is shown in *Figure 4.1*.

The surroundings are considered to be an infinite extended, isotropic medium with a far away temperature of zero Kelvin. T/T_k is given for sphere radii a = 10, 30, 50, and 75 m. The graphs illustrate the maximum temperature gradients which occur around the storage unit. The small size storage sphere clearly results in

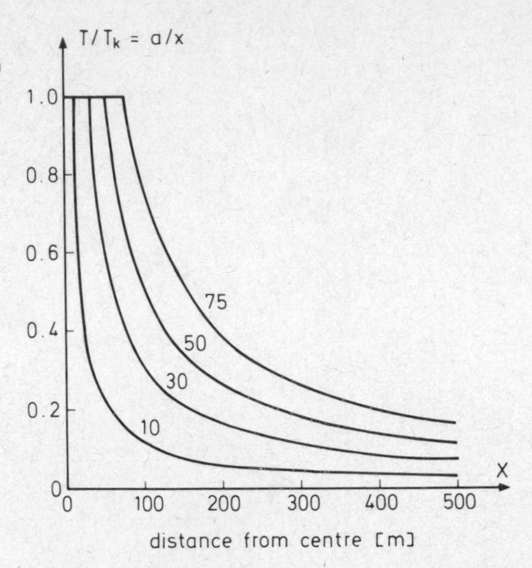

Figure 4.1
Temperature distribution around a storage sphere with fixed temperature T_k and placed in a isotropic infinite medium

the largest temperature gradient, hence relatively the largest heat loss to the surroundings.

The overall size of heat store is important and during recent years interest has been focused upon the possibility of using very large underground reservoirs for long term storage of heat on a community scale. Such units need relatively less insulation than a small storage unit for a single house.

Hot water

Energy can be stored as sensible heat by virtue of a rise in temperature of a material. For that purpose water is excellent, not only because of its low cost but also because of its high heat capacity (4180 J/kg/°C). Due to its low melting and boiling points, water is suitable as storage medium mainly between 5 °C and 95 °C.

Small water tanks are widely used for solar heat storage. A system of that kind is shown in principle in *Figure 4.2*. The usual

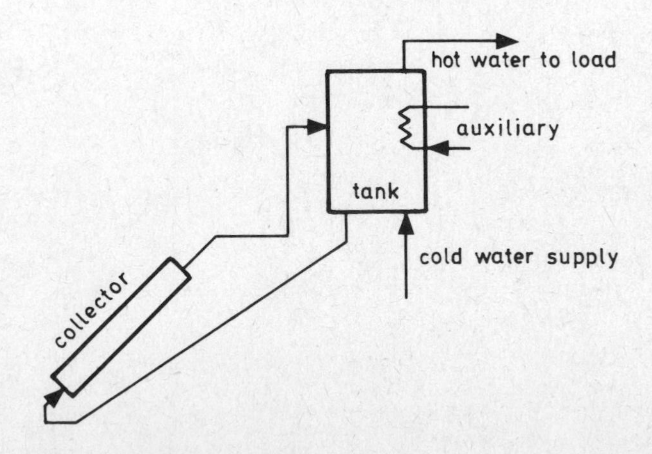

Figure 4.2
Hot water solar panel system with natural circulation

insulation used for individual house heating storage tanks is mineral or glass wool. Insulation has to be sufficient to provide for storage time of several months since load demand and available energy, as mentioned earlier, are out of phase, as shown in *Figure 4.3*. The optimum design for a single house solar heating system deals with the size of the storage unit in comparison with the area of the solar panel. In temperate locations, the optimum size of the storage unit is larger than in areas where yearly variations of solar energy are less.

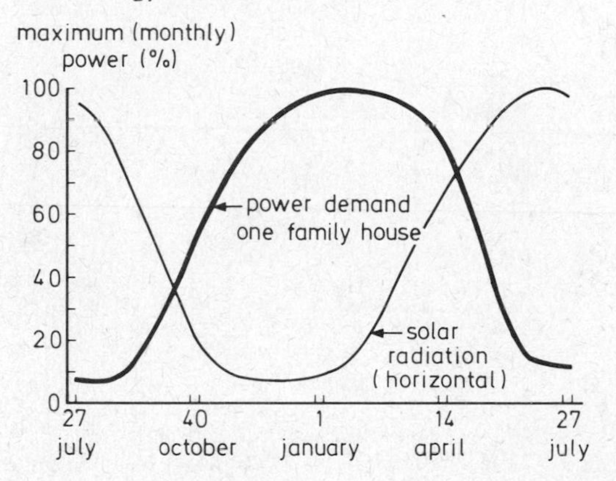

Figure 4.3
Yearly variation of heat demand and solar energy inlet for a northern European house (Lundén, A., Chalmers Tekniske Högskola, Sweden)

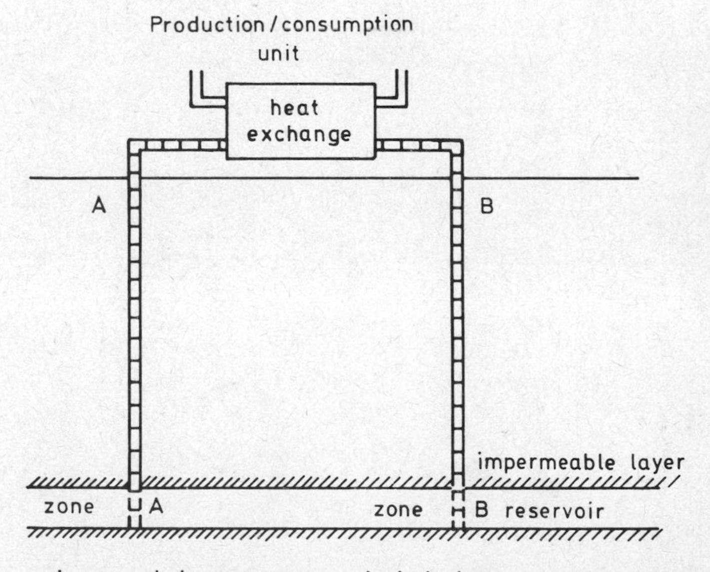

Figure 4.4
Underground layer heat storage (Technical University of Denmark)

Large scale hot water storage includes heat storage for combined heat and power plants (see Chapter 8) and for storage of heat energy for whole communities. Proposals for the latter mentioned application have included storage in surface lakes and in underground water bearing strata. Surface storage has associated problems such as the concern of fresh water supply and the risks of chemical and thermal pollution. Hence we will concentrate on the underground storage systems in the following, as shown in principle in *Figure 4.4*.

One of the essential parameters for such storage systems is the quantity of water which can be transferred between the underground storage and the surface. The water transfer is determined by the layer thickness h times the permeability k and the pressure gradient ΔP. This can be expressed by the flow equation

$$Q = p \frac{k h \Delta P}{\mu \ln r_1/r_2} \tag{4.1}$$

where

Q = amount of water per time unit
p = proportionality constant
k = permeability
h = layer thickness
ΔP = pressure difference between outer and inner storage
μ = viscosity of water
r_1 and r_2 = radius of the outer and inner storage limits.

If zone B is the storage zone for thermal energy, surplus heat from the production/consumption unit is used in the heat exchanger to heat water from zone A, which is returned heated to the storage stratum in zone B. In periods with heat deficit the flow is reversed and the store gives off heat.

Hot solids

Hot rocks and fire place bricks have served as heat storage from ancient times. This is still the case in industrial furnaces and in the baker's electric oven, where cheap electricity is used to heat the oven during the night.

High temperature heat storage is of interest both for utilising the heat in industrial processes and for heat engines. One recent example is power supply for Stirling engines. The energy used during a temperature change of say 50 °C is of the order of 10 Wh/kg for rocks, concrete, and iron ore. The volume density of the latter is double (~60 Wh/dm^3) that of the former because of the smaller mass density. The values for different solids are shown in *Appendix A*.

Applications for extremely high temperatures (>1000 °C) have been suggested, but it seems that materials problems such as corrosion, heat shocks and other problems associated with heat transfer have been prohibitive so far.

Phase change materials

Phase change materials offer a much larger heat capacity over a limited temperature range than sensible heat systems and can essentially supply heat at constant temperature. When heat is added to or removed from materials, phase change can occur in a variety of ways such as

melting
evaporation
lattice change
change of crystal-bound water content

where total energy change is given by the change in enthalpy.

Some inorganic salts, e.g. the fluorides (see *Appendix A*), exhibit large heats of fusion values but their high melting temperature causes severe corrosion problems. In order to lower the melting temperature the following eutectic mixtures have been investigated:

	melting point (°C)
Sodium-magnesium-fluoride, NaF/MgF_2	832
Lithium-magnesium-fluoride, LiF/MgF_2	746
Sodium-calcium-magnesium-fluoride, $NaF/CaF_2/MgF_2$	745
Lithium-sodium-magnesium-fluoride, $LiF/NaF/MgF_2$	632

The advantages of these fluoride mixtures are that they are chemically stable and can be contained in chromium nickel steel. Their phase change temperature may be convenient for use as storage for heat engines, but it is much too high for space heating systems.

Salt hydrates have suitable phase change temperatures for use as storage in heating systems. The phase transition, however, is often more complex than a simple melting, having a solid residue along with a dilute solution.

One of the salt hydrates most often proposed is Glauber salt, $Na_2SO_4 \cdot 10H_2O$. It decomposes at 32 °C to a saturated water solution of Na_2SO_4 plus an anhydrous residue of Na_2SO_4, the exchange of heat being 70 Wh/kg. The storage capacity per unit volume over a small temperature range is much larger than for water and since the civil engineering cost is the major expense for a house storage unit, salt hydrates may be more economical than water storage.

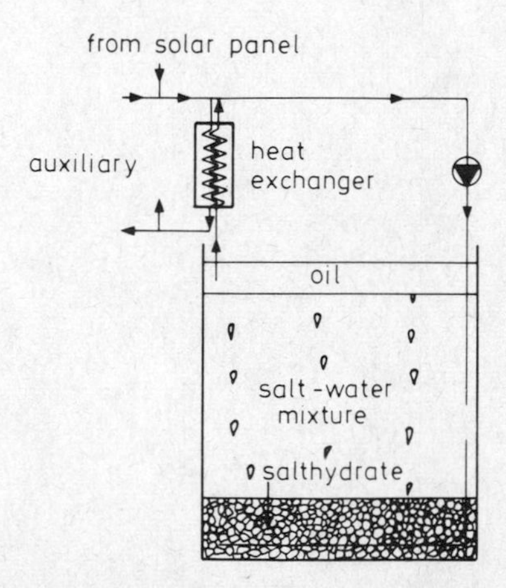

Figure 4.5
Solar panel heat storage with salt hydrate as a storage medium (Laboratory for Heat Insulation, Technical University of Denmark)

Figure 4.5 shows a solar panel heat storage with salt hydrates as a storage medium. The fluid through the heat exchanger is oil which is 'blown' into the lower part of the storage tank. The oil acts as a stirrer in the salt water mixture.

Metal hydrides have been suggested for thermal storages. The phase change is the absorption of hydrogen into the lattice of the metal or the metal alloy during a reaction of the following form:

metal + hydrogen $\rightleftarrows$ hydride + heat

This problem is looked at in more detail in Chapter 5, since it primarily deals with the problem of storing hydrogen in solids. It should be mentioned, however, that investigations, especially in USA, have proved that a combination of two hydrides in a so-called hybrid storage system is indeed applicable to heating/air conditioning systems. The advantage of this type of heat storage is that there are no losses during storage and the rate of heat reformation is easy to control. The main objective in hydride research is to find a cheap metal or metal alloy which works at suitable transition temperatures and pressures.

Before finishing this chapter, it should be mentioned that utilities around the world have shown recent interest in chemical reaction systems for thermal energy store. In particular, the following high temperature reaction has been studied.

$$CO + 3H_2 \rightleftarrows CH_4 + H_2O \qquad (4.2)$$

During off-peak hours, the heat from the primary source is absorbed in the reactor (reformer) where the previously stored reactants, methane and water, are converted into carbon monoxide and hydrogen. After heat exchange with the incoming reactants, the products are then stored in a separate vessel at ambient temperature conditions. Although the reverse reaction is thermodynamically favoured, it will not occur at these low temperatures and the storage time is in practice infinite. During peak hours the reverse reaction (methanation) is run and the resultant heat is used to generate electricity.

Further reading

Eggers Lura, A., 'Flat Plate Solar Collectors and Their Application to Dwellings', EEC Report 207–75–9, ECI DK (1976)

Furbo, S. and Svendsen, S., 'Heat Storage in a Solar Heating System using Salt Hydrates', Thermal Insulation Laboratory, Technical University of Denmark, Report 70 (1977)

Meyer, C. P. and Todd, D. K., 'Conserving Energy with Heat Storage Wells', *Environmental Science and Technology*, 7 (6), (1973), PP 512–516

Sabady, P. R., *The Solar House*, Newnes-Butterworths (1978)

Schulten, R., *et al.*, 'The High Temperature Reactor and Process Applications', Proc. Br. Nuclear Energy Soc. Int. Conf. (1974).

Sørensen, B., *Renewable Energy*, Academic Press (1979)

Telkes, M., 'Thermal Energy Storage', 10th IECEC, Newark, Delaware (1975), Pp. 111–115

5 Chemical Storage

Synthetic fuels

By synthetic fuels is meant substitutes for oil and natural gas manufactured from coal or from 'biological waste'. The substitutes are mainly for use as fuels for combustion engines in the transport sector. Such substitutes include synthetic crude oil (syncrude), methanol (CH_3OH), ethanol (C_2H_5OH), and methane (CH_4). Hydrogen is another possible substitute, but since the so-called hydrogen economy is a much broader concept, also including a variety of stationary applications, a special section of this chapter will be devoted to hydrogen.

The infrastructure of storage and distribution systems involving liquid synthetic fuels is not very different from that of the natural oil based liquid fuels. But some of the synthetic fuels are more toxic and corrosive. This imposes some handling (safety) problems and container materials problems. Such problems, however, are not ascribed to liquid fuels derived from syncrude, and present refinery installations can be used.

The manufacture of petroleum from coal can be accomplished by the following basic routes:

direct hydrogenation
carbonisation
solvent extraction – hydrogenation
Fisher–Tropsch synthesis.

At the moment the cost of petrol derived from syncrude is higher than that of petrol from natural oil, but countries with an abundant supply of coal close to the surface could probably become competitive in the near future. In South Africa the so-called SASOL plant using the Fisher–Tropsch synthesis claims to be only twice as expensive as the present production price of petrol.

Production of methanol from coal initially requires a conversion of coal into carbon oxides and hydrogen

$$C + H_2O \rightarrow CO + H_2 \qquad (5.1)$$

$$CO + H_2O \rightarrow CO_2 + H_2 \qquad (5.2)$$

$$CO + 2H_2 \rightarrow CH_3OH \qquad (5.3)$$

$$CO_2 + 3H_2 \rightarrow CH_3OH + H_2O \qquad (5.4)$$

The resulting mixtures of CO, CO_2 and H_2 can either be reacted as in the Lurgi process (50 atm, 250 °C) or in a fluidised catalyst bed reactor (Cu/Zn catalyst fluidised in liquid hydrocarbon).

A recent investigation by the United States Department of Energy shows that methanol can also be produced from biomass

$$C_xH_yO_z + 2(x - \frac{z}{2})H_2O \rightleftharpoons x\ CO_2 + \left[\frac{y}{2} + 2(x - \frac{z}{2})\right]H_2 \quad (5.5)$$

The process involves destructive distillation of organic solids in the presence of water, and the final step (Equation (5.4)) is catalysed by cobalt molybdate.

By upgrading the hydrogen content ethanol can be produced from CO and H_2 in the same way as methanol. Biomass is an alternative basic raw material attractive in areas of the world without coal. Ethanol can be produced from biomass by fermenting simple sugars to alcohol and carbon dioxide. The preliminary saccharification (hydrolysis) process to convert cellulose into monosaccharides involves either a combination of heat and acid or enzyme fermentation. The latter method has been chosen by a recent governmental research programme in Brazil. The Brazilian programme is aiming at a 20 % alcohol substitution to transport fuels in the early 1980s.

The advantages and disadvantages of the synthetic fuels are shown in *Table 5.1*.

Table 5.1 Summary of the most important advantages and disadvantages in using alternative fuels in Otto engines

	Advantages	*Disadvantages*
Methanol	'lean burn' lower consumption lower NO_x	cold start problems gives off a bad smell corrosion
Methyl fuel < 20 % methanol	'lean burn' lower consumption lower NO_x less lead	sensitive to water corrosion
Ethyl fuel < 25 % ethanol	less lead	sensitive to water
Methane	'lean burn' low pollution lower consumption	the tank system power reduction
Hydrogen	'lean burn' very low pollution	the tank system 'back-firing' power reduction (higher consumption)
Ammonia		gives off a bad smell toxic corrosion power reduction

Methanol has about half the volumetric energy density of petrol, it is very corrosive and has a high heat of vaporisation. The storage tanks in cars have to be double size and metal alloys currently used (also in engine blocks) have to be avoided because of corrosion. As an alternative fuel in internal combustion engines, it is probably best used as an extender of petrol, or as a source of on-board hydrogen.

Ethanol, which suffers from cold start problems and therefore requires manifold heating, is best used as a petrol extender.

Blending of methanol with petrol to act as an extender would obviate the need for alterations to conventional piston and Wankel engines. Reduction of emissions and knocking with lead-free petrol can be achieved. In addition it has been considered that methanol would make an excellent turbine fuel.

Ammonia is the least suited alternative fuel for use in internal combustion engines, but it has been suggested as a means of moving bulk energy. It can be stored in the same way as bottled gas. However, fittings of stainless steel have to be used instead of copper alloys in order to avoid corrosion. One interesting proposal is that ammonia could be produced by potential hydropower in Greenland and shipped in tankers to Europe. This would be a means of moving energy from an area of the world where no demand for bulk energy exists to an area with high demand. Ammonia could act as the transfer medium of energy from remote locations from where high voltage connections are impossible.

Methane is well suited for stationary bulk energy storage in underground salt caverns. The gas is stored under a pressure of 100–300 atm., depending on the depth and the natural pressure in the salt formation. Caverns in salt-domes are formed by irrigation of fresh water or salt water at a depth of about 1 km below the surface. 1000 cubic metres of gas at the above mentioned pressures require a volume of about 8–4 m^3 and the stored energy content is equivalent to 0.9 ton oil. Methane (or natural gas) storage technology is well developed and a considerable storage capacity has been established (see Chapter 8).

Hydrogen

Hydrogen is widely regarded as the ultimate fuel and energy storage medium for future centuries. This is because it can, in principle, be derived from water using any primary source of high quality energy, and it can be combusted back to water in a closed chemical cycle releasing no pollution. In addition the potential of hydrogen for the storage and cheap transmission of energy over long distances has led to the concept of 'Hydrogen Economy'. An energy distribution scheme is shown diagrammatically in *Figure 5.1*.

Hydrogen can be produced by several methods:

1. Catalytic steam reforming of natural gas
2. Partial oxidation of heavy oils

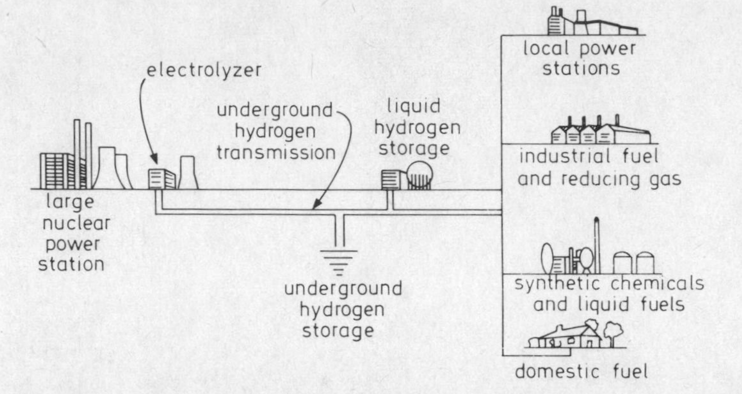

Figure 5.1
The hydrogen economy fuel system (Institute of Gas Technology, USA)

3. Water gas reaction, i.e. chemical reduction by means of coal following Equations (5.1) and (5.2)
4. Electrolytic decomposition of water
5. Biochemical, i.e. industrial photosynthesis
6. Ultraviolet radiation
7. Thermal decomposition of water, utilising thermochemical cycles.

Before the Second World War, most hydrogen was produced by reacting steam with heated coke following route (3), but today the routes (1) and (2) are predominant. Route (7) is subject to a large scale research effort at the EEC research centre at Ispra, Italy. It remains to be seen whether the problems of heat and mass transfer, heat recycling, solids handling, reaction kinetics, corrosion etc. can be overcome. Production of hydrogen is certainly not problem free.

By comparison with hydrogen production, its distribution is reasonably straightforward. One of the attractions of hydrogen as an energy vector is, indeed, that pipeline transmission over very long distances is cheaper and less objectionable on environmental grounds than electricity distribution.

We will not go into further detailed discussions about the problems involved in hydrogen production and distribution. The essential aim of this section is to review aspects of the storage of hydrogen. Two distinct aspects appear:

the bulk storage of large quantities of hydrogen produced from substantial non-oil based primary energy sources

dispersed storage of small quantities of hydrogen for transport applications.

Compressed gaseous hydrogen in underground caverns and liquid hydrogen in insulated containers seem to be possibilities for bulk storage. The cost of liquefaction is such that this means of storage is less attractive for stationary applications.

Hydrogen can be stored in underground systems in a way similar to natural gas because most rock structures are actually sealed by water in capillary pores. The high diffusivity of hydrogen gas has therefore little effect on the tendency to leak.

For small scale storage three possibilities exist:

compressed gas in cylinders
liquid hydrogen
metallic hydrides.

The mass and volume energy densities w_m and w_V are as follows:

	w_m (kJ/kg)	w_V (kJ/dm³)
Gas at 150 atm, 20 °C	140 000	1 700
Liquid, − 252 °C	140 000	10 500
Metal hydride (including metal)	1 400–11 000	17 500–21 000
Oil (for comparison)	44 000	40 000

Liquid hydrogen is particularly attractive to aircraft designers because of the three times better mass energy density, allowing the possibility of improved payloads and longer range, although it occupies about four times the volume of hydrocarbon fuels. A typical design published by Lockheed USA is shown in *Figure 5.2.*

Liquid hydrogen is the most probable hydrogen-fuel for heavy duty surface transport, i.e. large lorries and long distance motorway transport, because of its high mass energy density w_m. But interest in hydrogen-fuelled surface transport has so far been confined mostly to the private car and to urban transport vehicles. And here the hydride concept (i.e. hydrogen dissolved in metals), in particular, has been investigated.

Today there are definite projects for the development of a hydrogen-driven car based on metal hydride fuel (see Chapter 8). In principle, such cars operate by the release of hydrogen to the engine through heat supplied from the exhaust gas from the engine.

Work on metal hydrides has already made it clear that there are a number of systems, e.g. $LaNi_5$, TiFe, MgAl, and Mg_2Ni,

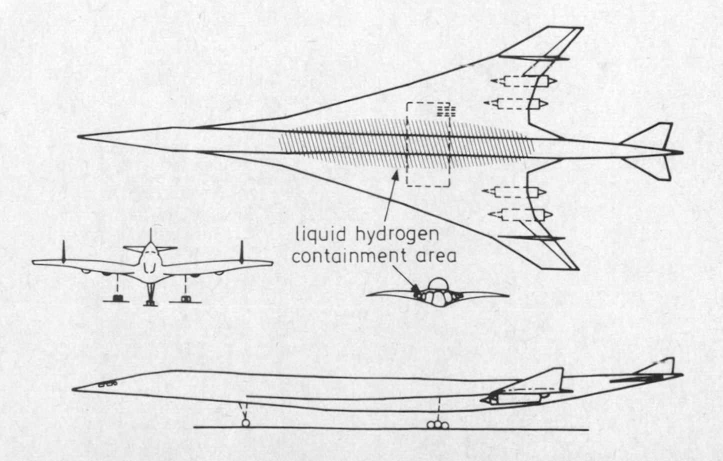

Figure 5.2
Suggested design for liquid hydrogen fuelled supersonic aeroplane (Lockheed, USA)

which can absorb hydrogen and release it at low pressures and temperatures with only slight losses.

The principal disadvantages of gaseous hydrogen as a storage medium is that it takes up very much space, it is explosive, and it is difficult to confine leak-free. The drawbacks with liquid hydrogen, aside from the inconvenience of such a highly cryogenic and inflammable liquid, are the sophisticated engineering required for production and transfer to and from the storage vessels and the cost of these operations.

These drawbacks of hydrogen are done away with in the hydride form; but since one rarely gets anything for nothing, it is hardly surprising that other problems spring up. The greatest ones are weight and the high price of 'host metals'. With regard to stationary energy stores, the price of the metal is the deciding factor, whereas in the case of transportation and use as fuel for vehicles, it is a combination of price and weight that counts.

The aim is to select a hydride which can be thermally decomposed in a reversible manner so that hydrogen may be withdrawn from the tank during use and replenished subsequently. Some of the desired features of a suitable hydride store besides low cost are

high hydrogen content per unit mass of metal
low dissociation pressure at easily accessable temperatures
constancy of dissociation pressure throughout decomposition
safe on exposure to air.

One storage material that has attracted special interest is FeTi hydride as developed by Brookhaven National Laboratories, USA. This is because it is a low temperature hydride with a low energy requirement for hydrogen release. Magnesium, which provides high temperature hydrides, has attracted interest because it is a readily available and therefore cheap metal. Both materials release hydrogen endothermically, thereby creating no safety problems. The mass energy densities (w_m) of hydrides based on Ti and Mg are:

$$FeTiH_{1.7} \rightarrow FeTiH_{0.1} \quad 516 \text{ Wh/kg} \tag{5.6}$$

$$Mg_2NiH_4 \rightarrow Mg_2NiH_{0.3} \quad 1121 \text{ Wh/kg} \tag{5.7}$$

$$MgH_2 \rightarrow MgH_{0.05} \quad 2555 \text{ Wh/kg} \tag{5.8}$$

The chemical reaction of exothermic hydride formation from metals (Me) and hydrogen (H_2) is as follows:

$$H_2 + Me \underset{\text{Discharging (heat added)}}{\overset{\text{Charging (heat released)}}{\rightleftarrows}} \text{hydride} + \text{heat} \tag{5.9}$$

If hydrides are used as hydrogen stores for heat engines or for domestic heaters the waste heat from such installations can be transferred back to the hydride. If the amount of waste heat is less than the heat needed for hydrogen release then the waste heat energy is stored in the metal hydride. The metal hydride acts as a heat store and a combination of hydrides with different

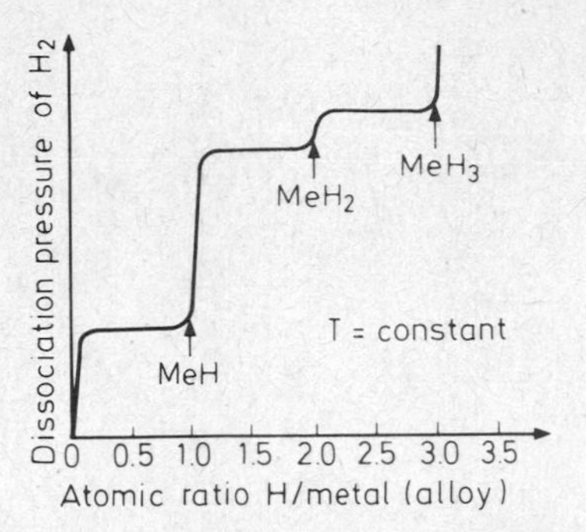

Figure 5.3
Pressure isotherms for different hydrogen concentrations in a hydride

release temperatures has a variety of applications such as heat pump systems, combined heating and air conditioning systems, and propulsion systems.

When molecular hydrogen interacts with a metal or an alloy, the hydrogen molecules are dissociated to atoms which are absorbed in the metal. When the limit of solution is reached an equilibrium hydrogen pressure is established which is characteristic of the composition. When all material has changed into hydride the pressure increases intensively again. If a second phase exists, another pressure level will arise and if there are no more phases the pressure increases steeply. An illustration of concentration pressure isotherms (plateau pressures) at a particular temperature is shown in *Figure 5.3*.

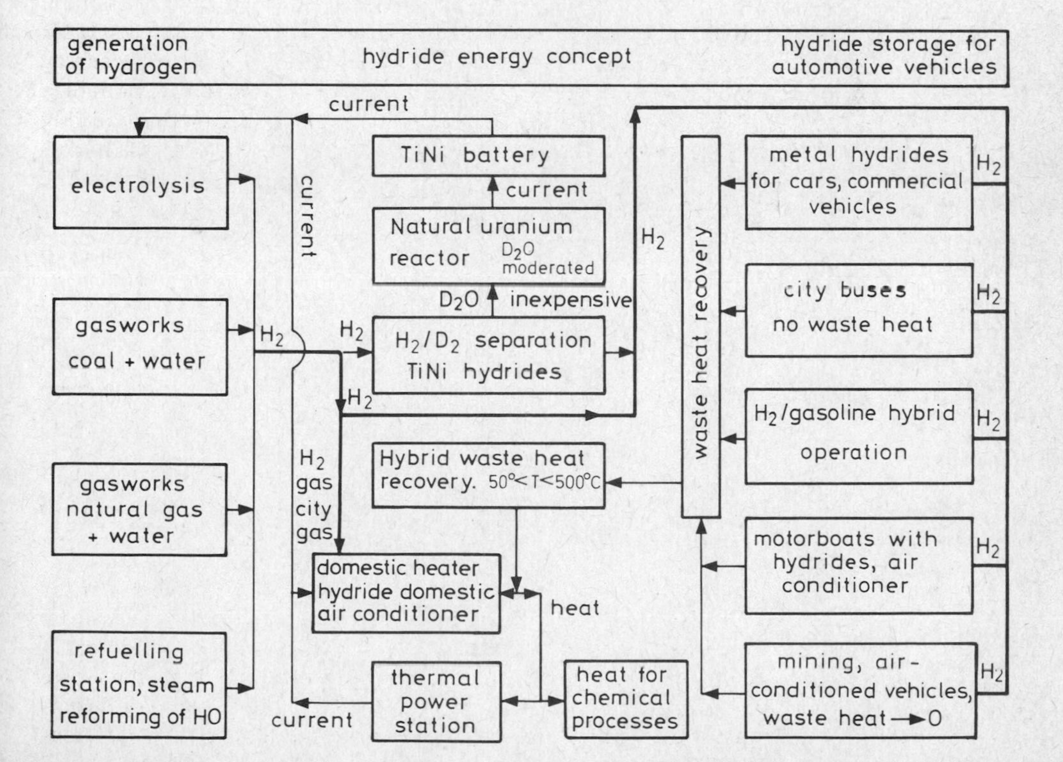

Figure 5.4
The Hydride Energy Concept (Daimler-Benz, West Germany)

The excess pressures at which hydrogen is released from the hydride are orders of magnitude less than that of hydrogen in the pressurised gaseous state because of the dissociation of molecules into atoms and the bonding in the metallic phases. For the same reason the volume density of hydrogen in hydrides is greater than in the gaseous and liquid form.

A combination of hydrides with different plateau pressures offers, as mentioned, a variety of applications involving both hydrogen storage and heat storage. This has lead to the so-called 'Hydride Concept' proposed by Daimler–Benz in Germany. *Figure 5.4* gives a survey of the Hydride Energy Concept, where hydrides can be used for:

mobile storage
stationary storage
electrochemical storage
hydrogen/deuterium separation.

The Hydride Energy Concept as developed by Daimler–Benz is based on the principle that a separation in time and location between the combustion process (in vehicle engines) and the release of the waste heat produced during this process is possible since

the temperature is determined by the pressure/temperature characteristics of the hydride

the heat transfer per unit of time can be controlled by varying the rate at which hydrogen is withdrawn or added from or to the hydride.

A combination of a hydride which releases hydrogen at low temperatures and a high temperature hydride works in the following way (see *Figure 5.5*). The exhaust gas from the hydrogen driven combustion engine passes the high temperature hydride, whereby hydrogen is released and fed into the low temperature hydride. The exhaust gas thereafter passes the low temperature hydride thereby releasing hydrogen to the engine and/or increasing the temperature of the hydride. Cooling water from the engine is also used for hydrogen dissociation. The end result is that the

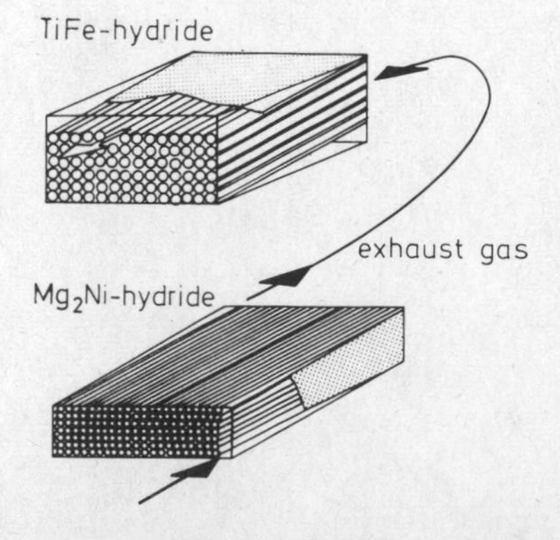

Figure 5.5
Hydride combination system for vehicle propulsion (Daimler-Benz, West Germany)

Table 5.2 Comparison of various fuels

	Petrol	*Methanol*	*Ethanol*	*Methane*	*Propane*	*Ammonia*	*Hydrogen*
Boiling point ω (K)	350–400	337	352	111.7	230.8	240	20.3
Liquid density (kg/m^3)	702	797	790	425	507	771	71
Heating value (mass) (kJ/kg)	44 380	20 100	27 700	50 000	46 400	18 600	120 000
Heating value (liquid) (MJ/m^3)	31 170	16 020	22 000	21 250	23 520	14 350	8 960
Diffusivity in air (cm^2/s)	0.08	0.16	–	0.20	0.10	0.20	0.63

hydride bed temperature reaches about 80 °C at the end of a more than 100 km test run.

The storage tank has to be cooled down to ambient temperature by means of external cooling water prior to being recharged. This heat release at the service station in addition to the heat released by refuelling (see Equation (5.9)) could be used as district heating in areas close to the service station. Such systems whereby waste heat is utilised for heating houses provide a considerable improvement of overall energy efficiency and as such means energy conservation, but because of the need for a complete change of the total infrastructure of vehicle design and energy supply in the transportation sector, a large scale replacement of present-day internal combustion engines by hydrogen powered engines seems impracticable for at least the next 20 years.

Also, large scale utilisation of hydrogen as fuel for stationary applications is widely regarded as rather futuristic. Today the bulk of hydrogen is produced from low cost oil and natural gas, and it is used almost exclusively for chemical purposes such as synthesis of ammonia, methanol, petrochemicals, and for hydrocracking within oil refineries. Hydrogen as fuel amounts to less than one percent of the yearly production and therefore it is difficult at present to define the cost of hydrogen for large scale distribution. It cannot at present, however, compete economically with fossil fuels and this situation is likely to continue until alternative primary energy sources become substantially cheaper than fossil fuels. Even then, there are major technical problems to be solved in the production, utilisation, and storage of hydrogen.

Electrochemical energy sources

Electrochemical energy sources can roughly be classified in the following categories: primary batteries, secondary batteries, and fuel cells. A common feature of these energy sources is that chemical energy is converted to electric energy. The efficiency for this process is not Carnot limited, as opposed to thermal processes. Primary and secondary batteries utilise the chemicals built into them, whereas fuel cells have chemically bound energy supplied from the outside in the form of 'fuel', e.g. hydrogen, methanol, or hydrazine. Contrary to secondary batteries, primary batteries cannot be recharged when the built-in active chemicals have been used. An important parameter of batteries for traction as well as for load levelling is the number of charges and discharges possible. The term 'batteries' will refer to secondary batteries in the following text.

Batteries and fuel cells consist of two electrode systems fitted on each side of an electrolyte. The electrodes exchange ions with the electrolyte and electrons with the outer circuit. The two electrodes are called (during discharge) the anode (–) and cathode (+) respectively (see *Figure 5.6*).

The anode is defined as the oxidising electrode, i.e. the electrode sending positive ions into the electrolyte during discharge (and

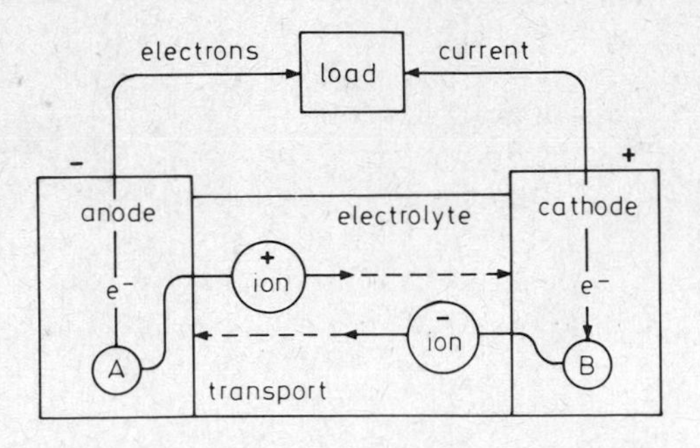

Figure 5.6
An electrochemical energy source during discharge

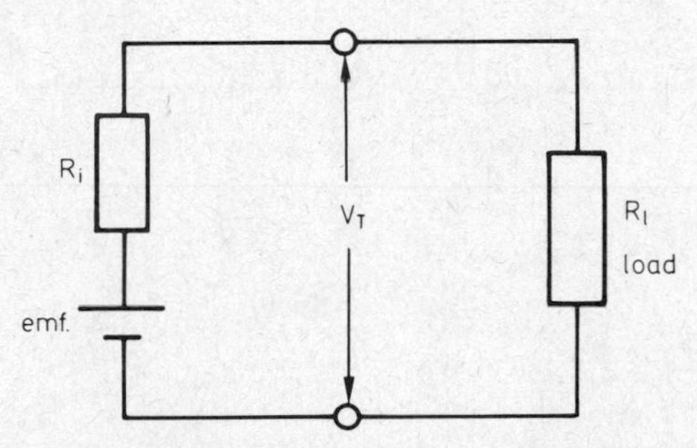

Figure 5.7
Simplified circuit diagram

electrons to outer circuit). By supplying positive current to the electrolyte phase, the anode itself becomes negatively charged and acts as electron source for the outer circuit. The cathode acts as electron drain. The electromotive force of the batteries (emf) equals the difference between the electric potential of the electrodes. The difference in potential constitutes the motive force for the electric current. The terminal voltage V_T equals the electromotive force minus the voltage drop in the battery, and in order to maintain the terminal voltage and consequently the current, electrons have to be produced at the anode and used at the cathode. As no chemical process can produce a net charge, a transport of charge in the electrolyte in the form of ions between the electrodes has to take place.

The electrolyte must conduct ions, but no electron conduction must take place, as in that case the cell will 'short-circuit' and discharge itself. In a battery the mobile positive ions may be metal ions, produced by the anode metal supplying electrons to the outer circuit. In a fuel cell it may be H^+ (H_3O^+), produced by the hydrogen supplied to the anode. A very simplified transport model is shown in *Figure 5.7.*

The inner resistance R_i contains frequency and time dependent components associated with the electrode processes, as well as an ohmic resistance against the charge transport in the entire inner circuit. R_i measurements shows that it is dependent on the load as well as on the remaining energy contents of the battery. Hence the inner resistance can only be described in the form of a rather complicated impedance. In conclusion it can be said that a resistance as small as possible is needed in order to decrease the losses in the system.

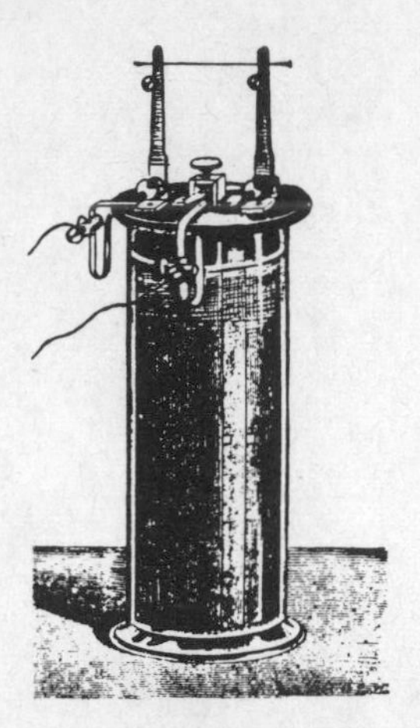

Figure 5.8
The lead battery invented by Planté in 1859

To decrease the irreversible losses in an electrochemical system, it is important to attain a high reaction rate and good transport conditions. Both factors can be met by working at high temperatures and with very active electrode systems. In both cases the electrolyte will often prove to be a limiting factor with regard to stability and transport properties.

Aqueous electrolytes can work at high temperatures only when under pressure. A possible solution, however, lies in the use of ceramic materials, having a suitably high specific conductivity for ions, which can take part in the electrochemical process. The development of such materials, called solid state ion conductors, has contributed to a break-through of electrochemical energy storage and conversion.

Batteries

Battery manufacture is still dominated by the lead battery invented by Plante in 1859 (see *Figure 5.8*). Industrial development on alkaline electrolyte batteries (such as Ni–Zn, Fe–Ni) aims to produce improved power systems for vehicular transport in the medium future. Advanced batteries, in particular the Na/S couple using a solid state electrolyte, and the Li/S couple using a fused salt electrolyte, are under development in several centres.

A comparison of energy densities shows that the advanced systems are attractive:

System	*Theoretical Energy Density* (Wh/kg)	*Expected Realisable Density* (Wh/kg)
Lead–acid	167	40
Fe–Ni	266	60
Ni-Zn	321	90
Na/S	680	150
Li/S	1500	150

Nevertheless, in spite of the expenditure of several millions of dollars over more than ten years, no commercial Na/S or Li/S battery is yet in production. Also while the next two to three years should show whether such a product is feasible, there remain several potential disadvantages apart from the necessity for high temperature operation:

Na/S electrolyte optimisation (conductivity/composition/ strength)

electrolyte breakdown under charge containment

safety aspects of liquid sodium at 350 °C.

Li/S electrodes
separators
safety aspects of liquid lithium at 400 °C.

The main problems lie in designing, fabricating and optimising materials and, additionally, battery design involves interactions between several scientific disciplines including ceramics, solid state chemistry, electrochemistry, metallurgy, and engineering.

In particular it would be desirable to lower the temperature of operation, raise the energy density and to make an inherently safe system while maintaining low overall cost.

Results on alternative battery concepts indicate that such improvements may be possible, and particular interest applies to ambient temperature organic electrolyte batteries using solid solution electrodes (e.g. Li/TiS_2 Exxon Corp.), all solid state batteries involving composite electrolytes (e.g. Li(Si)/LiI, Al_2O_3/ TiS_2, Mallory) or solid polymer electrolytes. Additionally, the high

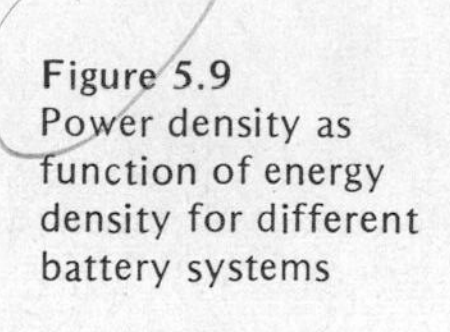

Figure 5.9
Power density as function of energy density for different battery systems

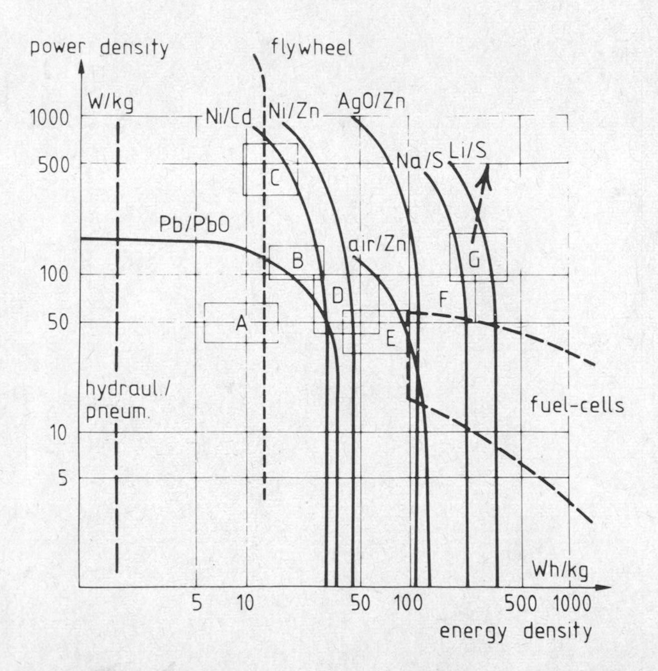

surface area of such systems can entail lower current densities with consequent improvement in reliability.

The purely technical demands that can be made on any energy system for use in the field of traction are

sufficiently high energy density
power density
longevity.

If these demands can be met, the transport industry would be interested in a new traction system with view to further investigations of the most important parameters for this industry, i.e. reliability and cost.

A number of investigations have been carried out to define the sufficiently high energy density and power density of an electrochemical system. An example of this is *Figure 5.9*, which was taken from *Proceedings – Electric Vehicles Study Days*, International Union of Producers and Distributors of Electrical Energy, March 1972. *Figure 5.9* shows the power density as function of energy density for different battery systems. By way of comparison values for hydraulic/pneumatic storages, flywheels and fuel cells have been added.

The energy density determines the range and the power density the acceleration capacity of a vehicle. The squares in *Figure 5.9* A–G denote the demands made on various types of vehicles. It should be noted, however, that demand F – a normal family car – probably has been underrated in relation to the performance of a car with a petrol engine.

In *Table 5.3* are shown some candidate electric vehicle batteries and in *Table 5.4* a list of candidate batteries for load levelling as judged by the Electric Power Research Institute (EPRI), USA, from their study in 1976.

As mentioned previously, the lead-acid battery has been the popular choice as power source both for traction and stationary purposes. The main drawbacks are low energy density, slow recharging time, and the need for careful maintenance.

The cell reaction is as follows:

$$\underset{\substack{\ominus \\ \text{Anode}}}{Pb} + 2\,H_2SO_4 + \underset{\substack{\oplus \\ \text{Cathode}}}{PbO_2} \underset{\text{Charge}}{\overset{\text{Discharge}}{\rightleftarrows}} \underset{\substack{\oplus \\ \text{Anode}}}{PbSO_4} + 2\,H_2O + \underset{\substack{\ominus \\ \text{Cathode}}}{PbSO_4} \qquad (5.10)$$

During discharge the cathode is positive and the anode negative. About half the weight of a lead–acid battery is accounted for by inert materials, e.g. grid metal, water, separators, connectors, terminals and cell containers, and attempts to reduce the weight have involved the use of low-density grid materials. The introduction of carbon fibres into positive electrode grids has resulted not only in reduction in weight, but also in increased power capability of lead–acid cells.

Table 5.3 Candidate electric vehicle batteries (Rand, D.A.J., 1977)

	Battery				Current Performance		
System	*Electrolyte*	*Temp.* (°C)	*O.C.P.* (V)	*Energy Density (Theoretical)* (Wh/kg)	*Energy Density* (Wh/kg)	*Power (Peak)* (W/kg)	*Cycle Life*
Lead–acid	H_2SO_4	20–30	2.05	171	22[a]	50	700+
Lead–acid (Improved)	H_2SO_4	20–30	2.05	171	30[a]	–	500+
Ni–Zn	KOH	20–30	1.706	321	66[b]	150+	400+[c]
Ni–Zn (Vibrocel)	KOH	20–30	1.706	321	45–65[b]	–	1 200+
Ni–Fe	KOH	20–30	1.370	267	60[b], 82.5[d]	50–100	1 500+
Fe–air	KOH	~40	1.280	764	81[d]	30–40	200+
Zn–air	KOH (stagnant)	50–60	1.645	1 084	131.5[d]	–	200–300
Zn–air	KOH (flowing)	50–60	1.645	1 084	109[d]	26	200–300
Zn–air	KOH (slurry)	50–55	1.645	1 084	110[e]	80[e]	500–600
$Zn–Cl_2 . 6H_2O$	$ZnCl_2$	50	2.12	465	110–157[f]	88–132[f]	400+
Na–S	β-Al_2O_3	300–375	1.76–2.08	664	180[f]	220[f]	300+
Li/Al–FeS_2	LiCl-NCl (eutectic)	400–450	1.5	625	70[b][c]	50[c]	250+
Li/Al–FeS	LiCl-KCl (eutectic)	400–450	1.6	869	40–60[c]	–	600+
Li–TiS_2	?	90–110	1.87–2.5	480	132	132	120+

(a) At 1 hour rate *(b) At 2 hour rate* *(c) Cell only*
(d) At 5 hour rate *(e) Projected figures* *(f) At 4 hour rate*

Table 5.4 Load-levelling batteries: candidates and characteristics. (EPRI Journal Vol. 1, 8, 1976)

	Operating temperature (°C)	*Theoretical cell energy density* (Wh/lb)	*Design cell energy density* (Wh/lb)	*Design modular volumetric energy density* (Wh/in^3)
Lead–acid (Pb/PbO_2)	20–30	110	9	0.75
Sodium–sulfur (Na/S)	300–350	360	70	2.5
Sodium–antimony trichloride ($Na/SbCl_3$)	200	350	50	2.0
Lithium–metal sulfide ($LiSi/FeS_2$)	400–450	430	85	3.5
Zinc–chlorine (Zn/Cl_2)	50	210	25	0.7

In the following, the cell reactions for some of the other battery systems will be listed together with a brief description of major associated problem areas.

The nickel–zinc battery is analogous to the much more expensive nickel-cadmium battery. The cell reaction is:

$$\overset{\oplus}{2\,NiOOH} + 2\,H_2O + \overset{\ominus}{Zn} \underset{\text{Charge}}{\overset{\text{Discharge}}{\rightleftharpoons}} 2\,Ni(OH)_2 + Zn(OH)_2 \quad (5.11)$$

The main problem with nickel–zinc is its short cycle life. Other problems include cost, mass production problems, separator stability and temperature control. Poor cycle life is caused by the high solubility of the reaction products at the zinc electrodes. Redeposition of zinc during charging results in the growth of dendrites (which penetrate the separators of the battery and cause an internal short circuit) and also in the redistribution of active material. Attempts have been made to suppress the growth of zinc dendrites during charging by vibrating the zinc electrode.

The nickel–iron battery is an alkaline storage battery using KOH as the electrolyte. The cell reaction is:

$$\overset{\oplus}{2\,NiOOH} + 2\,H_2O + \overset{\ominus}{Fe} \underset{\text{Charge}}{\overset{\text{Discharge}}{\rightleftharpoons}} 2\,Ni(OH)_2 + Fe(OH)_2 \quad (5.12)$$

The major objection to nickel–iron batteries for electric vehicle applications has been their mediocre energy density. However, recent developments with Japanese and Swedish systems are fast placing this battery on a par with the nickel–zinc system. The battery does, however, suffer from a poor peaking capability. Other shortcomings of the nickel–iron battery are low cell voltage (necessitating more cells for a given battery voltage) and the low hydrogen overvoltage of the iron electrode which results in self-discharge and low cell efficiency.

Depth of discharge[a] (%)	*Density – 10 h rate* (mA/cm²)	*Operating potential* (V)	*Demonstrated cell size* (kWh)	*Demonstrated cell life* (cycles)	*Critical materials*
25	10–15	1.9	> 20	> 2000	Lead
85	75	1.8	0.5	400	None
80–90	25	2.6	0.02	175	Antimony
80	30	1.4	1.0	1000	Lithium
100	40–50	1.9	1.7	100	Ruthenium (catalyst)

(a) Also known as utilisation of active materials

The iron–air battery consists of an anode using iron as active material and a cathode taking oxygen from the air. The cell reaction is:

$$\overset{\ominus}{Fe} + H_2O + \overset{\oplus}{\tfrac{1}{2} O_2} \underset{\text{Charge}}{\overset{\text{Discharge}}{\rightleftharpoons}} Fe(OH)_2 \qquad (5.13)$$

The battery suffers from high self-discharge of the iron electrode at low temperatures, poor charge efficiency and limited power capacity (max 30–40 W/kg).

The zinc–air battery has a highly concentrated KOH electrolyte and the electrochemical reaction is between oxygen from the air and zinc metal. The cell reaction can be written as:

$$\overset{\ominus}{Zn} + \overset{\oplus}{\tfrac{1}{2} O_2} \underset{\text{Charge}}{\overset{\text{Discharge}}{\rightleftharpoons}} ZnO \qquad (5.14)$$

As with the iron–air system, the zinc–air couple has a poor overall charge-discharge efficiency due to the polarization losses associated with the air electrode. The main difficulty in developing a secondary system lies in the instability of the zinc electrode. The zinc oxide formed during discharge dissolves in the electrolyte to give zincate ions and redeposition of zinc during charging leads to the problems of electrode shape change and dendritic growth discussed above for nickel–zinc batteries. Attempts to produce even zinc deposits by the high speed circulation of electrolyte have not resulted in any significant improvement in cell lifetime or performance.

The zinc-chlorine battery attempts to overcome the replating problem of zinc by using an acid electrolyte. The cell reaction is:

$$\overset{\ominus}{Zn} + \overset{\oplus}{Cl_2} . 6\,H_2O \underset{\text{Charge}}{\overset{\text{Discharge}}{\rightleftharpoons}} ZnCl_2 + 6\,H_2O \qquad (5.15)$$

The main problem with the zinc-chlorine hydrate battery is that the system is complex. The auxiliary equipment for refrigerating, heating and storing the chlorine places severe weight and volume penalties on the system.

The sodium–sulphur battery has a ceramic electrolyte (β-alumina) which can conduct sodium ions. The cell reaction can be written:

$$\overset{\ominus}{x\,Na} + \overset{\oplus}{y\,S} \underset{\text{Charge}}{\overset{\text{Discharge}}{\rightleftharpoons}} Na_xS_y \qquad (5.16)$$

The problems of developing beta-alumina electrolytes for use in sodium–sulphur batteries include scientific problems of understanding how the electrical, mechanical, and chemical properties of beta-alumina depend upon its composition, phase equilibria and micro structure. A second problem concerns the technology of how to manufacture electrolyte tubes to the desired specifications. A third problem is related to overall design of the cells. Despite some progress, much remains to be done, especially in the field of ceramic production technology.

The lithium–sulphur battery consists of liquid lithium and sulphur electrodes and an electrolyte of molten LiCl–KCl eutectic at an operating temperature in the 380–450 °C range. The cell reaction is:

$$\overset{\ominus}{2\,Li} + \overset{\oplus}{S} \underset{\text{Charge}}{\overset{\text{Discharge}}{\rightleftharpoons}} Li_2S \qquad (5.17)$$

Problems encountered include attack on ceramic insulators and separators by the highly corrosive liquid lithium and problems with selfdischarge caused by lithium dissolving in the molten LiCl–KCl electrolyte.

The use of lithium–aluminium alloys and iron–sulphide electrodes has resulted in development of more practical Li–S cells with reasonable energy densities. The reactions are:

$$\overset{\ominus}{4\,LiAl} + \overset{\oplus}{FeS_2} \underset{\text{Charge}}{\overset{\text{Discharge}}{\rightleftharpoons}} Fe + 2\,Li_2S + 4\,Al \qquad (5.18)$$

and

$$\overset{\ominus}{2\,LiAl} + \overset{\oplus}{FeS} \underset{\text{Charge}}{\overset{\text{Discharge}}{\rightleftharpoons}} Fe + Li_2S + 2\,Al \qquad (5.19)$$

The lithium–titanium disulphide battery has an intervalation cathode of TiS_2 and a lithium metal anode. The electrochemical discharge reaction proceeds via the insertion of lithium ions between adjacent sulphur layers. The reaction is:

$$\overset{\ominus}{x\,Li} + \overset{\oplus}{TiS_2} \underset{\text{Charge}}{\overset{\text{Discharge}}{\rightleftarrows}} Li_xTiS_2 \qquad (5.20)$$

The intercalation reaction occurs with no change in the host matrix except for a slight expansion of the *c*-axis. The charge/discharge process shows good reversibility and operates at room temperature. The battery which is not yet commercially available has a number of outstandint features. These include high energy density, long life, a leak-proof seal and the ability to indicate the level of its charge.

Redox cells such as the rechargeable flow cell $TiCl_3 | TiCl_4 \| FeCl_3 | FeCl_2$ have been considered as bulk storage systems for utility use. This cell consists of two electrolyte compartments and inert carbon electrodes separated by a selective ion exchange membrane. On discharge, $FeCl_3$ is reduced to $FeCl_2$, while $TiCl_3$ is oxidised to $TiCl_4$. The membrane allows the passage of chlorine ions (Cl^-) between the electrolyte compartments to preserve electroneutrality. The redox flow cell operates at ambient temperature and the overall efficiency is 70%. Unlike conventional batteries, there are no apparent cycle life limitations due to changes in the active electrode materials. However there is considerable uncertainty about the total costs of the system.

At this stage, a more detailed description of beta-alumina and the sodium–sulphur battery is felt to be appropriate, because this is the advanced battery that has been under development for the longest time.

Advanced batteries have provided the stimulus for much interesting work in solid state chemistry and electrochemistry in recent years. The seminal discovery was the observation of very high sodium ion mobility above room temperature in sodium beta-alumina by Weber and Kummer. Beta-alumina, one of a family of closely related compounds, has a layer structure as shown in *Figure 5.10*. Four layers of close packed oxygen atoms contain fourfold and sixfold co-ordinated aluminium in an atomic arrangement analogous to that found in spinel, $MgAl_2O_4$. The formula of beta-alumina, however, $Na_{1+x}Al_{11}O_{17+\frac{1}{2}x}$ is not compatible with an infinitely extending spinel structure. The relative lack of aluminium is reflected in the fact that every fifth layer of oxygen atoms is only one quarter filled. The sodium ions are found in this relatively empty layer.

Beta-alumina was originally so named because it was thought to be an isomorph of aluminium oxide, and the crystal structure was determined before the Second World War. However, the unusual electrical properties of sodium beta-alumina were not

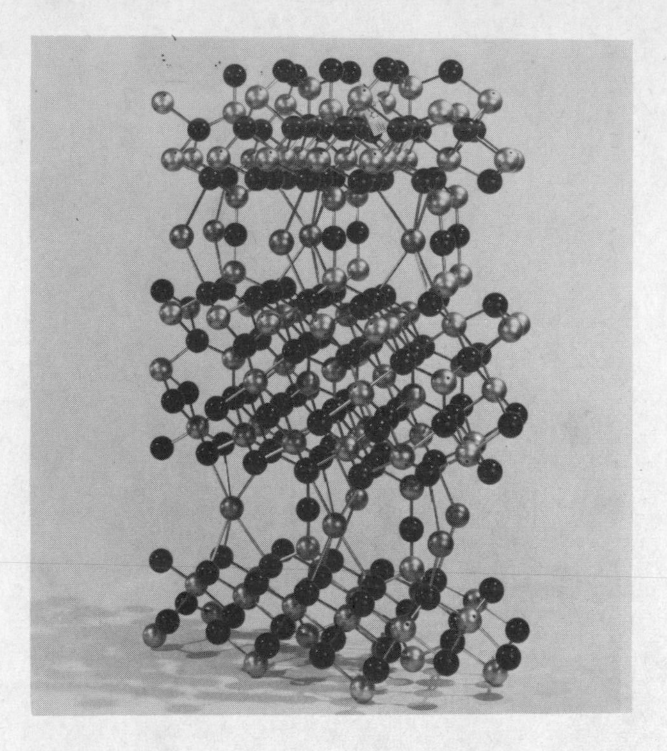

Figure 5.10
The beta alumina structure. The densely populated close-packed 'spinel blocks' containing aluminium and oxygen ions separate the 'mirror planes' which contain the mobile sodium ions. The density of oxygen atoms in the mirror planes is only one quarter of that in the spinel blocks. The separation between mirror planes is just over 11 Å. (AERE, Harwell, UK)

discovered until much more recently. When the electrical conductivity at 300 °C was measured, and it was realised that the current carriers were exclusively sodium ions and not electrons, the application of materials like beta-alumina to a modern generation of power batteries was quickly realised. In the sodium–sulphur battery, patented by Ford, instead of solid electrodes separated by a liquid electrolyte, as in the conventional lead–acid car battery for example, sodium beta-alumina is used as a solid electrolyte, specifically conducting sodium ions, between liquid electrodes of sodium metal and sulphur (*Figure 5.11*). The cell voltage, 2.08 V, is derived from the chemical reaction between sodium and sulphur to produce sodium polysulphide and the theoretical energy density, about 750 Wh/kg, is strikingly high when compared with that of the lead–acid battery, only about 170 Wh/kg.

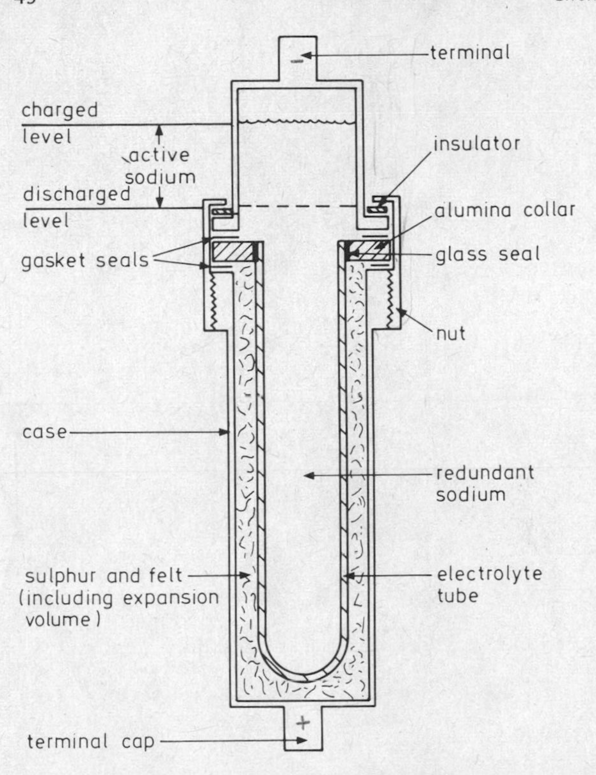

Figure 5.11
A schematic model of the sodium-sulphur battery which uses a sodium beta alumina solid electrolyte as the separator between liquid electrodes (sodium anode and sulphur cathode). The operating temperature is 300–400 °C (AERE, Harwell, UK)

Fuel cells

Most of the battery research throughout the world is concerned with materials research for advanced batteries, in particular materials for solid solution electrodes and for solid electrolytes. In this connection it should be noted that solid electrolytes are applicable in a wide range of electrochemical applications apart from the secondary batteries.

Consider an electrochemical concentration cell

electrode 1, $X(a_1)$ | solid electrolyte conducting X as an ion | $X(a_2)$, electrode 2

The emf of such a cell is given by the Nernst equation:

$$E = \frac{RT}{nF} \ln \frac{a_2}{a_1} \tag{5.21}$$

where n is the number of electrons needed to get one atom or molecule of X into its ionic form in the electrolyte
a_1 and a_2 are the activities at electrode 1 and 2.

We can make use of such a cell in the following ways:

1. If $a_1 \gg a_2$ and X is continually added on the left, and removed on the right we have a source of energy, a concentration fuel-cell.
2. If T and one a is known we can measure the other a. We have an electrochemical sensor.
3. If we apply a greater voltage than E in the opposite sense we can drive X from one side to the other. Hence we have an ion pump or an electrolyser.

Fuel cells, as distinguished from other secondary systems by their external fuel store, date back even further than the lead–acid battery. The first hydrogen–oxygen fuel cell was in principle demonstrated by the English lawyer W. R. Grove in 1839 (see *Figure 5.12*). But the bulk of the fuel cell's development lies in the last 30 years, and the major application has been in the space industry.

The contribution by a number of scientists such as O. K. Davtyan, G. H. Briers, J. A. A. Ketelaar, H. H. Chambers, A. D. S. Tantram, F. T. Bacon, K. V. Kordesch, and A. Marko during the 1950s has made it possible to arrive at technology for practical operating cells with different fuels and operating temperatures.

As a fuel cell is an electrochemical cell which can continuously change the chemical energy of a fuel and oxidant to electric energy with high efficiency, it is not surprising that a variety of possible fuels has been tried, including hydrogen, methanol, ammonia, hydrazine, and methane. *Figure 5.13* shows a diagram of a hydrogen–oxygen cell.

Also a number of different electrolytes have been tested including solid state, acid, alkaline and molten carbonate. Although expectations were high regarding the high temperature ZrO_2

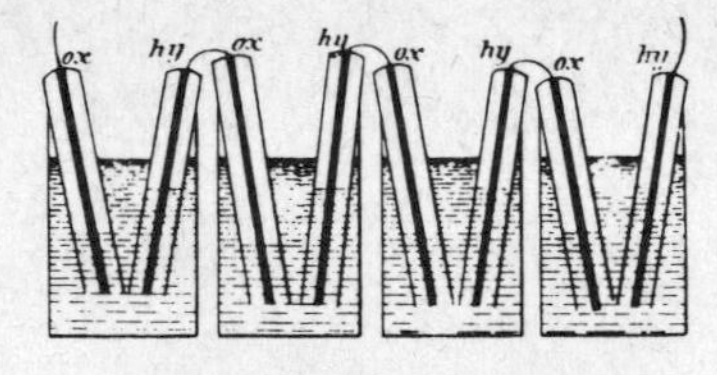

Figure 5.12
Fuel-cell by W. R. Grove, 1839

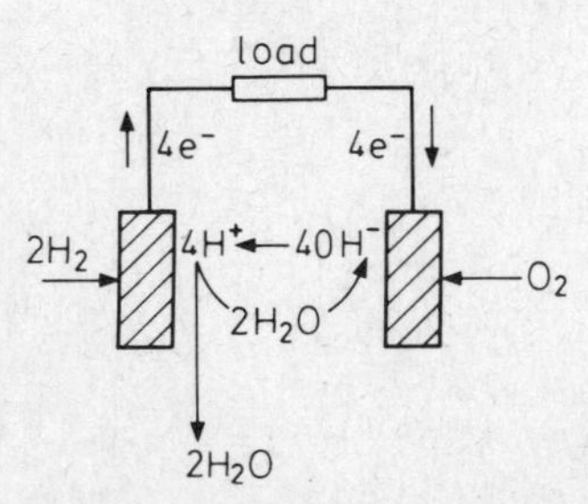

Figure 5.13
A hydrogen–oxygen fuel-cell

electrolyte systems, most research on this system has been cancelled because of severe corrosion problems. Two options seem promising at the moment: the phosphoric acid cell working at 160–200 °C and the molten carbonate cells working at 600–750 °C. The latter option is the advanced concept, where at the moment problems are paramount.

Hydrogen–oxygen fuel cells have attracted special interest mainly because of the previously mentioned concept of a hydrogen economy.

These fuel-cells with the overall reaction

$$H_2 + \frac{1}{2}O_2 \rightarrow H_2O \tag{5.22}$$

are attractive for reasons such as high energy density, no pollution and high cell efficiency. As seen from the overall equation a good hydrogen, or oxygen, conductor is the obvious type of electrolyte. Known systems with acid or alkaline electrolyte for low temperature operation, high or very high temperature melts and very high temperature solid electrolytes all exhibit major drawbacks.

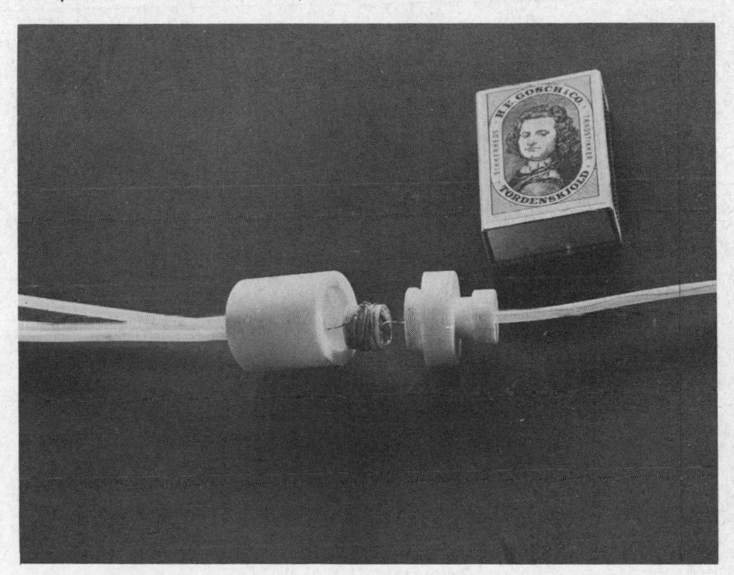

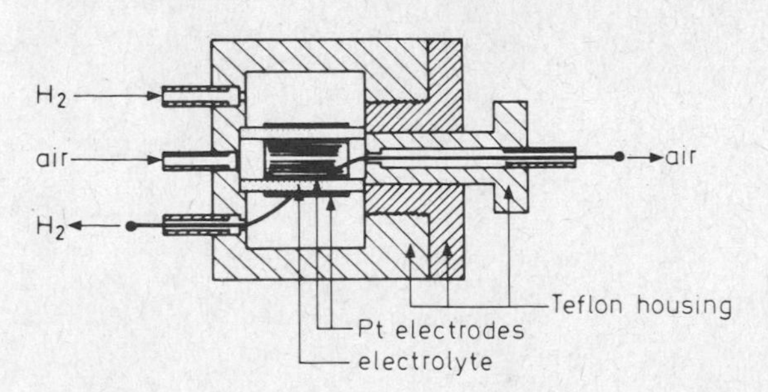

Figure 5.14
Fuel-cell with hydrogen exchanged beta-alumina as electrolyte (Dept. of Chemistry, Odense University, Denmark)

Attempts therefore have been made to produce a good hydrogen conducting solid electrolyte for medium temperature operation. Research in this field has been going on in Europe, USA and Japan for the last three years. One of the first medium temperature laboratory cells with a solid hydrogen conducting electrolyte was demonstrated early 1975 as a result of cooperative work between Harwell Laboratory, UK and Odense University, Denmark (see *Figure 5.14*).

The cell has a partially hydrogen exchanged beta-alumina electrolyte with Pt sputtered gas electrodes (~ 1500 Å). The exchange is performed by applying an electric field across the electrolyte with wet hydrogen flowing at the positive electrode. Fully hydrogen exchanged beta-alumina tubes are unfortunately not obtainable, and current research involves other routes such as alternative primary exchange materials and direct fabrication of solid hydrogen conductors.

As society becomes increasingly more dependent on electricity, generated from both coal and nuclear base load plants, and also from alternative sources such as wind, waves and the sun's rays, there will be an increasing demand for energy storage to match supply with demand. Batteries have some excellent properties in this respect:

they store and give up electric energy
being modular, they can be used flexibly
they are largely free of environmental problems
they can typically have a short lead time in manufacture.

Furthermore, the prospect of efficient off-peak energy utilisation by relatively non-polluting electric vehicles is likely to play an increasingly important part in the total picture. In both cases, the extent to which this can be achieved depends critically on the development of cheap, long-lifetime, high energy batteries to provide the performance capabilities which will be demanded.

In summary therefore advanced batteries and fuel cells can be seen to fulfil three roles in the field of efficient energy utilisation in the medium to long term future:

within the electrical supply system
smoothing of supply to demand for alternative energy sources
electric traction.

Additional applications of large scale industrial importance, such as traditional stand-by and emergency power systems, nautical and local storage systems, should also offer the prospect of increased battery use. By considering the above markets, in which battery storage of energy could lead to reduced dependence on fossil fuels and energy conservation, attempts are made in research throughout the world to arrive at reasonable estimates of market penetration of batteries in the future, judged against their technical performance and cost targets.

Battery storage for solar electricity is one of the fast developing application areas. In most present applications a rechargeable chemical battery is associated with the solar cells in order to bridge periods of insufficient sunshine. The solar cell-battery research has been increased in recent years because of the expected decrease in production cost for mass produced solar cells (see *Figure 5.15*).

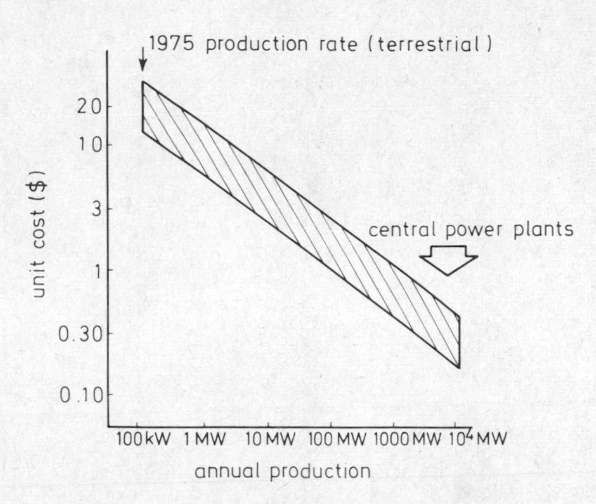

Figure 5.15 Projected learning curve for photovoltaic systems (Palz, W., 1978)

As solar cell costs fall, greater use of solar electricity will become directly dependent on storage costs, making cheap, long-life batteries essential to power generation by this method.

Further reading

Bruchner, H. 'The Hydrogen/Hydride Concept', *Int. Symp. on Hydrides for Energy Storage*, Institutt for Atomenergi, Kjeller, Norway (1977)

Cochran, N. P. 'Oil and Gas From Coal' *Scientific American*, Vol. 234, (1976)

Dell, R. M. and Bridger, N. J., 'Hydrogen – The Ultimate Fuel', *Applied Energy*, 1, (1975)

Jensen, J., 'Advanced Battery Development', *International Power Generation*, Vol. 2, No. 2 (1979)

Jensen, J., McGeehin, P. and Dell, R. M., 'Electric Batteries for Energy Storage and Conservation', Odense University Press (1979)

Kalhammer, F., 'Storage Batteries, the Case and the Candidates' *EPRI Journal*, Vol. 1 (1976)

Kordesch, K. V., '25 years of Fuel Cell Development (1951–1976)', *J. Electrochem. Soc.*, Vol. 125 (1978)

Palz, W., *Solar Electricity*, Butterworths, London (1978)

Rand, D. A. J., 'Current International Action on Advanced Battery Systems for Electric Vehicles', *J. Power Sources*, Vol. 4 No. 1 (1979)

Tofield, B. C., Dell, R. M. and Jensen, J., 'Advanced Batteries', *Nature*, Nov. 16 (1978)

Weyss, N., 'Wasserstoff – Informationen zum Wasserstoff Koncept, *Energie*, Jahrg. 26, (1974)

6 Mechanical Storage

Potential energy: springs, compressed gas, and pumped hydro storage

When a mechanical system is in a state where it has the potential by itself to release energy and thereby carry out work it is said to have 'potential energy'. An example is a mass m placed in the gravitational field g at a height h. Due to the gravitational force, the mass can move a distance Δh to a lower position h_0. If m is coupled to an external system the work that can be carried out is

$$W = \int F \, ds = m g \Delta h \qquad (6.1)$$

There are many historic examples of elevated weights used as energy stores, especially where a constant force is required. One example is the weight powered file-cutter designed by Leonardo da Vinci. *Figure 6.1* shows another simple example of potential mechanical energy storage. A weight with mass m is coupled to a door. When the door is opened energy is supplied in order to lift m to the higher position h. Energy has been supplied to the weight system, which now has the potential to close the door again. The same function is obtained by a mechanical compressible device where, by closing the door, the device would change its compression ratio. Only a small amount of stored energy is necessary to close a door and a small spring could certainly do the job.

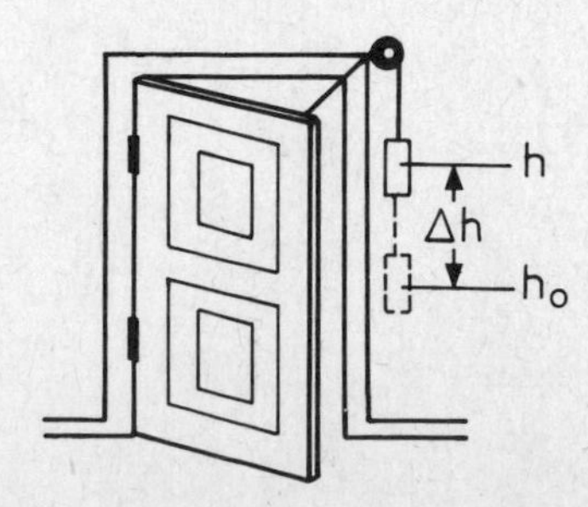

Figure 6.1
A weight-operated door closer

Springs

Springs are widely used to store small amounts of energy in mechanical positioning systems. Most applications, however, are found in dynamic mechanical systems where the primary purpose is not to store energy. One exception is the children's spring powered toys.

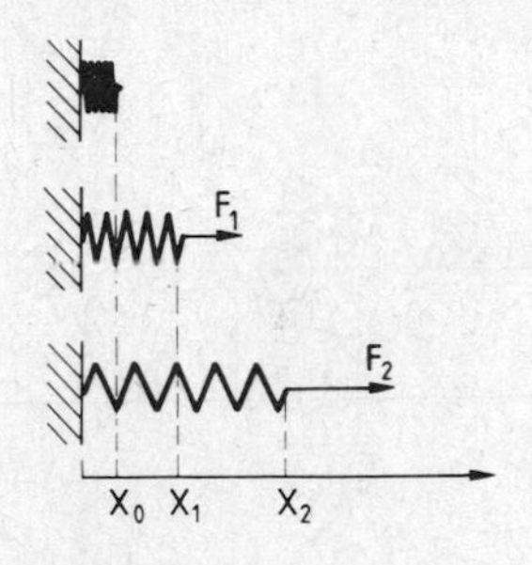

Figure 6.2
Expansion of a spring by applying forces of different values

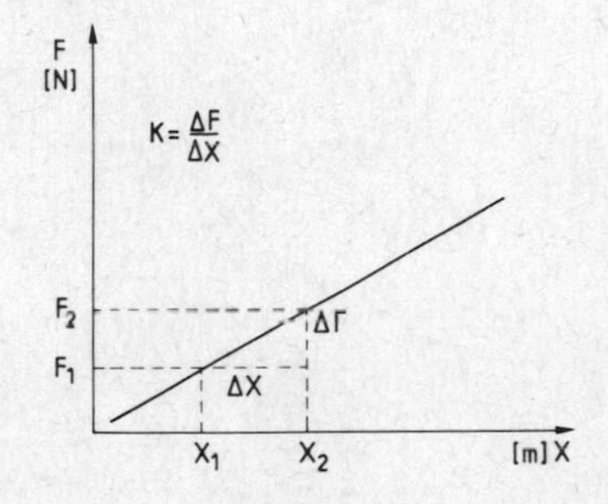

Figure 6.3
Distance of expansion as a function of force

The calculation of the amount of energy stored in a spring is of the same kind whether it relates to a compressed or an expanded spring. We will do the calculation of stored energy related to the latter as shown in *Figure 6.2*. By applying an external force F, the spring is expanded a distance x. An ideal spring exhibits a linear relationship between F and x as shown in *Figure 6.3*. The spring constant or elasticity constant is defined as

$$k = \frac{\Delta F}{\Delta x} = \frac{F}{s} \tag{6.2}$$

where s is the total expansion.

The work supplied in order to extend the spring to a distance of s can be derived by inserting $F = ks$ from Equation (6.2).

$$W = \int F\mathrm{d}s = \int ks\mathrm{d}s = \tfrac{1}{2}ks^2 \tag{6.3}$$

The energy density of a spring system or of any system with solid elastic material as the energy storage medium is rather low, whereas the power density is high. The energy and power density for steel springs and natural rubber without transfer mechanism are listed below.

	Energy density (Wh/kg)	*Power density (W/kg)*
Steel spring	0.1	10^4
Natural rubber	8	80

The high power density is dependent on how the energy is removed from the elastic system. In any case, the relatively small distances of compression or expansion result in large forces which again mean heavy transfer mechanics and low energy density. Because of the low energy density it is unlikely that elastic systems will find applications for large scale energy storage.

Compressed gas

Compressed gas is another way to obtain mechanical energy storage. When a piston is used to compress a gas, energy is stored in the gas and can be released later by reversing the movement of the piston. Pressurised gas is therefore an energy store. It can release energy which can be used to perform useful work.

The ideal gas law relates the pressure p, the volume V, and the temperature T of the gas as follows:

$$pV = nRT \tag{6.4}$$

where

p = pressure
V = volume
n = number of moles
R = gas constant

The amount of work that can be extracted from the compressed gas depends on the type of process applied. Apart from the purely adiabatic process, which does not occur in real processes since it would require infinitely large insulation, some degree of heat exchange is always associated with the process. The amount of heat follows from the first law of thermodynamics

$$\Delta U = Q + W \tag{6.5}$$

which states that the change in internal energy ΔU is equal to the sum of the heat Q and the useful work W.

In *Figure 6.4* is shown a cylinder with compressed gas at a pressure P. If we assume that the piston can move without friction we can now derive the work done by the compressed gas when the piston is forced to move a distance s to the right. We

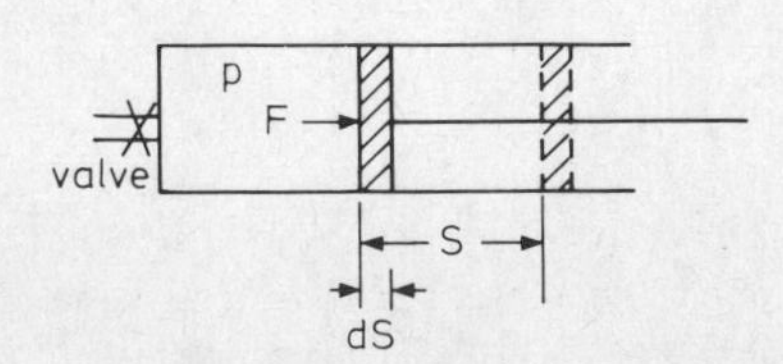

Figure 6.4
Gas at a pressure p moves a piston by applying a force F

get the force F on the piston with area A from the definition of pressure p

$$p = \frac{F}{A} \tag{6.6}$$

hence

$$W = \int F\,\mathrm{d}s = \int p\,A\,\mathrm{d}s = \int p\,\mathrm{d}V \tag{6.7}$$

This integral would be easy to solve if p was constant during the process, and for this purpose we could think of the valve in *Figure 6.4* being open to a very large pressure container. The very small change in total volume would not cause the pressure to drop and we would essentially have an isobar process where the work done by the gas is

$$W = p \int_{V_1}^{V_2} \mathrm{d}V = p\,(V_2 - V_1) \tag{6.8}$$

Here $V_2 - V_1$ is the volume traversed by the moving piston. Movement of the piston to the left (against the gas by applying an external force) stores a corresponding energy $p\,(V_2 - V_1)$.

If the process follows an isotherm instead of an isobar, what is then the solution of the integral in Equation (6.7)? In the case of constant temperature the gas law (Equation (6.4)) now becomes Boyle–Mariotte's law

$$pV = n\,R\,T = \text{constant} \tag{6.9}$$

hence

$$W = \int_{V_1}^{V_2} p\,\mathrm{d}V = n\,R\,T \int_{V_1}^{V_2} \frac{\mathrm{d}V}{V} = n\,R\,T\, \ln \frac{V_2}{V_1} \tag{6.10}$$

As a matter of interest, let us calculate the order of magnitude of the volumetric energy density for such a system. Our example could be a cylinder with a starting volume V_0 of 1 m^3 and a pressure p_0 of 2 atm. (one atm. = 101 325 Pa). If the gas is compressed to a volume of 0.4 m^3 at constant temperature the amount of stored energy is

$$W = n\,R\,T \int_{V_0}^{V} \frac{\mathrm{d}V}{V} = P_0\,V_0 \ln \frac{V_0}{V} = 1.86 \times 10^5 \text{ J}$$

Since we are dealing with a total volume of 1 m^3 the volumetric energy density in joules per m^3 is just the same number. The energy density is, as we shall see in Chapter 7, much higher than those of magnetic or electric fields.

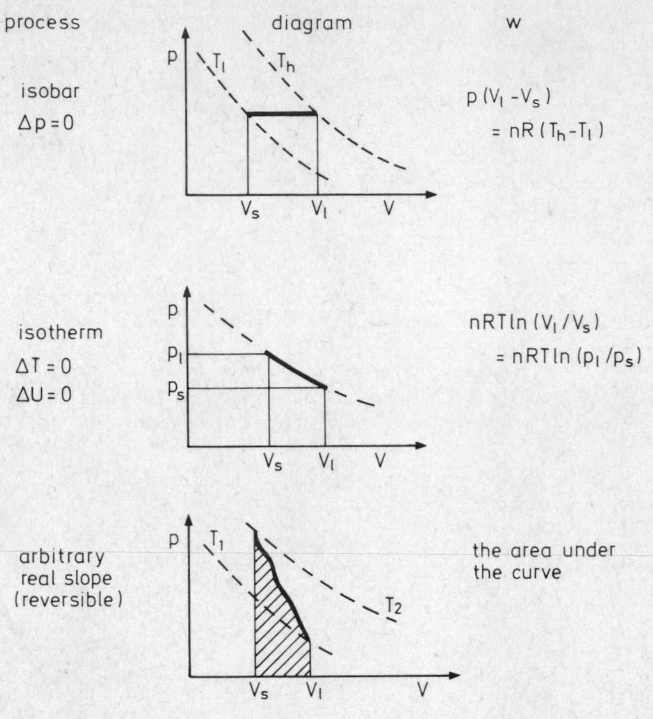

Figure 6.5
Processes with *n* mole of an ideal gas

Real processes do not maintain a constant pressure or a constant temperature exactly, and perhaps the third process mentioned in *Figure 6.5* is the most interesting to look at more closely. The third process is an arbitrary reversible one. *W* can be determined by the area under the curve no matter what slope is between the two volumes V_{small} and V_{large} and the corresponding temperatures T_{high} and T_{low}. It is not surprising that it is so because after all that is just what the general expression for *W* in Equation (6.7) tells us.

Large scale energy storage in the form of compressed air stored in natural or man-made underground caverns is economical and technically viable at present. The energy stored in the compressed air can be utilised using low pressure turbines, but this approach results in very low round-trip efficiencies. The alternative is to expand the air through a combustor. The advantage of the latter approach (see *Figure 6.6*) is that the turbine does not need to drive its compressor, thus increasing its output power and decreasing the fuel consumption.

Combined compressed air storage — gas turbine systems result in substantially lower generation cost for utility peaking power. They utilise proven and existing gas turbine technology and require minimal above-ground real estate.

Small scale storage with compressed gas, usually nitrogen, as the storage medium has been widely used for hydraulic systems.

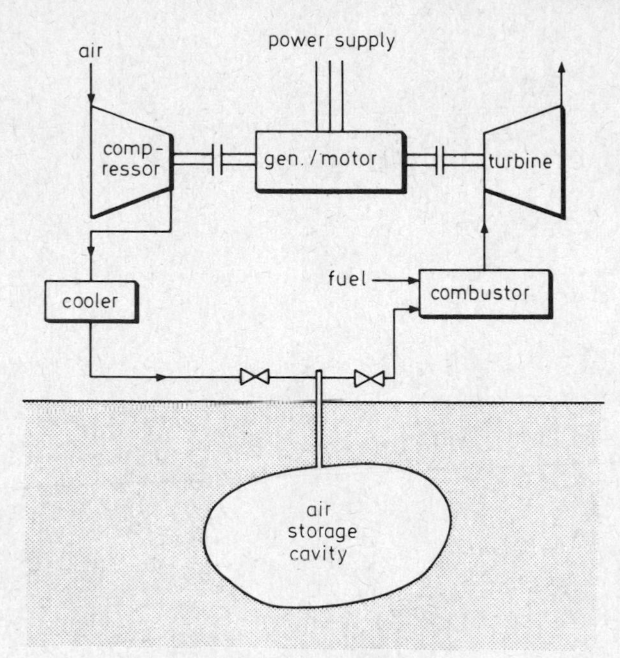

Figure 6.6
Compressed air storage

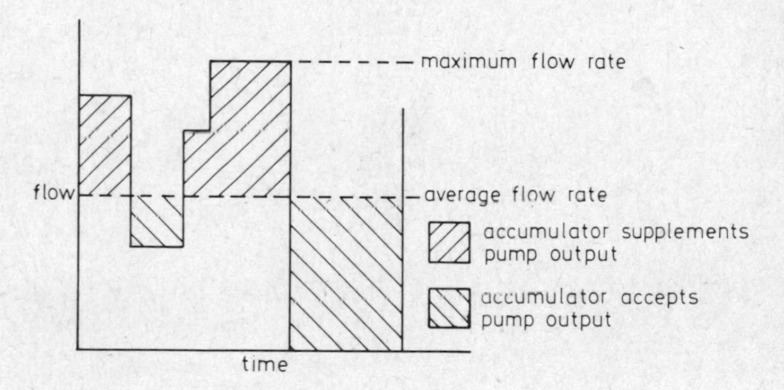

Figure 6.7
Typically operating cycle (Christie Hydraulics Ltd, UK)

Most hydraulic systems require variable and intermittent flow rates. Energy can be saved by using the storage device to accept pump output when system demand is low, and supplement output when demand is high.

Figure 6.7 shows a typical hydraulic system, repetitive, operating cycle, where the flow requirement varies over a period of time. Without an accumulator the pump and driving motor is sized to provide the maximum flow rate, with a resultant energy waste when demand is less than maximum. Now, if an accumulator is used, the pump and driving motor may be reduced in size to provide only the average flow rate over the time cycle.

There are additional benefits: other system components such as the oil reservoir, filter and certain valves may be reduced in size, thus lowering the capital cost; noise and heat generation may be reduced; finally, the system will show a quicker response to demand.

Most accumulator designs are based on the principle that gas is compressible and oil is nearly incompressible. Assume an inert gas, such as nitrogen, is contained under pressure in a vessel. If hydraulic fluid is pumped into that vessel at a higher pressure than that of the original gas, the nitrogen will compress, with pressure rising to that of the fluid being pumped. This increase in gas pressure is proportional to the decrease in volume.

The vessel now contains energy in that a volume of hydraulic fluid is stored under pressure against the compressibility of the nitrogen gas, and if the fluid is released from the vessel, it will be forced quickly out, under pressure, by the expanding gas. Accumulator design usually demands some means of separating the gas from the fluid in the pressure vessel, and the most popular method is to use a rubber bag or bladder (see *Figure 6.8*).

As well as the energy saving aspect for accumulators, there are many other useful storage applications: quicker system response to demand; counterbalance of heavy moving weights; leakage compensation. Shock elimination, impact absorption, cavitation elimination and pulse smoothing, are all helped by the use of the bladder accumulator, which has low inertia and friction.

Figure 6.8
Accumulator design with rubber bag for separating the gas from the fluid (Fawsett Engineering Ltd, UK) 1 Air valve protective cap, 2 Air valve stem locknut, 3 Separator bag, 4 Shell, 5 Poppet valve, 6 Fluid port locknut, 7 Fluid port assembly, 8 Bleed plug, 9 Anti-extrusion ring, 10 Distance piece, 11 'O' ring seal, 12 Gland ring, 13 Air valve, 14 Sealing cap

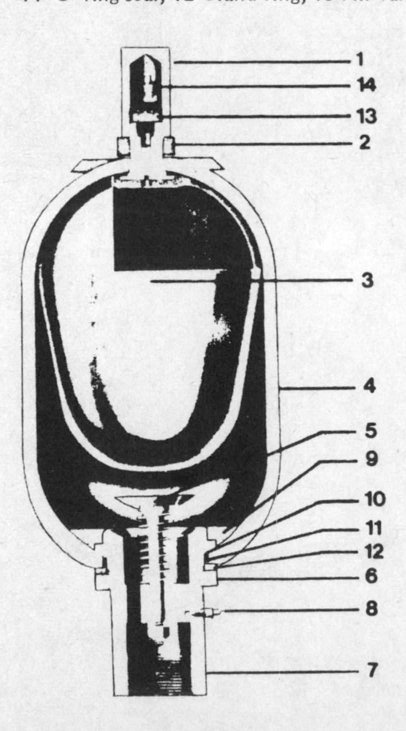

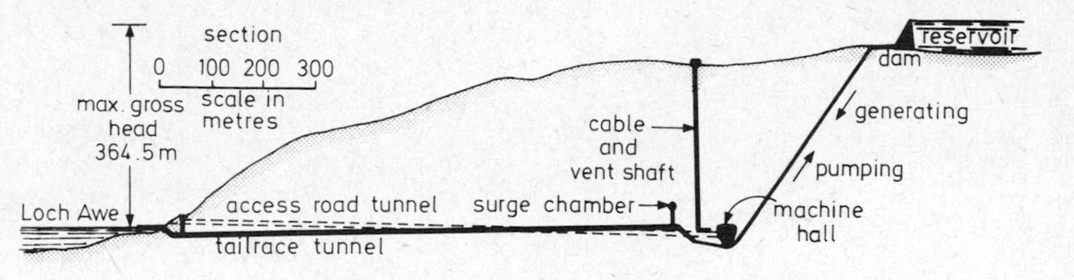

Figure 6.9
The North of Scotland Hydro-Electric Board Crauchan station (North of Scotland Hydro-electric Board, UK)

Pressure may also be transferred from one dissimilar fluid or gas to another, without contact.

Pumped hydro

Pumped hydro storage is the only large scale energy storage method in general use. For decades, utilities have used pumped hydro storage as an economical means of utilising off peak energy by pumping water to a reservoir at a higher level. During peak load periods the stored water is released through the pumps, now acting as turbines, to generate electricity to meet the peak demand.

One example of such an installation is the North of Scotland Hydro-Electric Board Cruachan station at Loch Awe (see *Figure 6.9*). It is described in more detail in Chapter 8. Here we will calculate the energy that can be released when water from the upper reservoir accelerates, turning most of its potential energy to kinetic energy.

Each litre or kilogram of water elevated 100 m has a potential energy of

$$W_{pot} = m\,g\,h = 1 \times 9.8 \times 100 = 980 \text{ J/kg} \tag{6.10}$$

The upper limit v_{max} for the speed of the water as it enters the turbines can be deduced by equating the kinetic energy and the potential energy, assuming zero losses.

$$W_{kin} = W_{pot} \tag{6.11}$$

$$\tfrac{1}{2}\, m\, v^2{}_{max} = W_{pot} \tag{6.12}$$

hence

$$v_{max} = 2\, W_{pot} = 2 \times 980 = 44.3 \text{ m/s} \tag{6.13}$$

In the case of the Cruachan station where the difference in elevation is 370 m the corresponding figures are

$$W_{pot} = 3630 \text{ J/kg} \qquad v_{max} = 85 \text{ m/s}$$

In operation there are losses such as frictional losses in turbulence and viscous drag, and the turbine itself is not a perfect machine. The water retains some kinetic energy as it enters the tailrace. For the final conversion of power to electricity we also have to account for losses in the generators. We define the overall efficiency as the ratio of useful power extracted in the conversion to the actual power released. In the installation at Gruachan this efficiency is near to 90 % and as a rule pumped hydro installations exhibit high efficiencies ranging from 80–90 %.

Kinetic energy: flywheels

In inertial energy storage systems, energy is stored in the rotating mass of a flywheel. This way of storing energy for shorter periods of time is one of the oldest known (see *Figure 6.10*). In ancient potteries, the kick at the lower wheel of the rotating table was the energy input to maintain the rotation. The rotating mass stores the short energy input so that rotation can be maintained at a fairly constant speed. Flywheels for just the same purpose have been applied in steam and combustion engines since the time of development of these engines. The application of flywheels for longer time storage is much more recent (see Chapter 8). It has been made possible by development in materials science and in bearing technology.

The energy content of a rotating mechanical system is:

$$W = \tfrac{1}{2} I \omega^2 \qquad (6.14)$$

where

I = moment of inertia
ω = angular velocity

The moment of inertia is determined by the mass and the shape of the flywheel. I is defined as follows:

$$I = \int x^2 \, dm_x \qquad (6.15)$$

where x is the distance from the axis of rotation to the differential mass element dm_x. If we consider a flywheel of radius r in which the mass is concentrated in the rim (see *Figure 6.11*), then the solution of the integral (6.15) is simple since $x = r =$ constant.

$$I = x^2 \int dm_x = r^2 m$$

and

$$W = \tfrac{1}{2} r^2 m \omega^2 \qquad (6.16)$$

Equation (6.16) tells us that the energy content depends on the total mass to the first power and on the angular velocity (or the number of revolutions per time unit) to the second power. This means that in order to obtain high energy content, high angular velocity is much more important than the total mass of the rotary system.

The energy density w_m, i.e. the amount of energy per kg, can be derived directly by dividing by m

$$w_m = \tfrac{1}{2} r^2 \omega^2 \qquad (6.17)$$

Figure 6.10
Ancient pottery. The flywheel is used to maintain constant speed of the pottery table

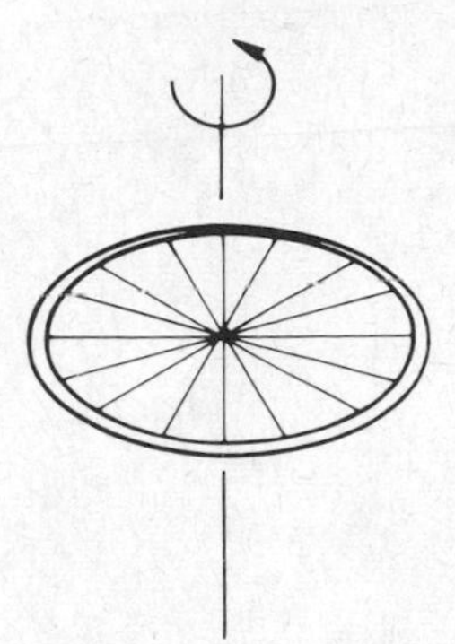

Figure 6.11
Flywheel with mass concentrated in the rim

The volume energy density w_{Vol} is derived by substituting m in Equation (6.16) with m expressed as the mass density ρ multiplied by the volume.

$$w_{Vol} = \tfrac{1}{2}\,\rho\, r^2\, \omega^2 \tag{6.18}$$

High angular velocity depends on the strength of the material. In our example the tensile stress σ in the rim is given by

$$\sigma = \rho\, \omega^2\, r^2 \tag{6.19}$$

so that the maximum kinetic energy per unit volume will be

$$w_{Vol}\ (\text{maximum}) = \tfrac{1}{2}\,\sigma_{max} \tag{6.20}$$

Thus if the dimensions of the flywheel are fixed, the main requirement is high tensile strength. High values of w_{Vol} are important in some cases, but in the case of transport applications it would clearly be better to use the maximum w_m as a criterion. By combining Equations (6.17), (6.18) and (6.20), w_m as a function of ρ and σ is

$$w_m\ (\text{maximum}) = \tfrac{1}{2}\,\frac{\sigma_{max}}{\rho} \tag{6.21}$$

so that a light material with a high tensile strength is required. Fibre composites with strengths higher than steel and much lower mass densities are an obvious choice. The tensile strength and mass densities of some fibres which might be considered are given in *Table 6.1*.

Table 6.1

Fibre	*Strength* (GN/m²)	*Mass density* (g/cm³)
Glass	3.5	2.5
Silica	6.0	2.2
Carbon	2.6	1.9
Chrysolite asbestos	4.5	2.5

For comparison the value of strength for steel wire (0.9 % carbon) is 4.2 but the mass density is rather high (7.9).

The factor 0.5 in Equation (6.21) relates to the chosen example of a simple rim flywheel, but the expression is valid for any flywheel, made from material of uniform mass density ρ.

$$w_m \text{ (maximum)} = c\frac{\sigma_{max}}{\rho} \tag{6.22}$$

The value of c depends on the geometry of the flywheel, and c is therefore usually called the shape factor or the form factor. Essentially the c value comes from the expression of the moment of inertia I. Values of I for different shapes of rotating bodies are shown in *Figure 6.12*.

In order to obtain maximum energy storage density a special design, where maximum stress throughout the flywheel is obtained, has been proposed. Such flywheels are thickest near the axis, and thinnest near the rim, in contrast to the classical flywheel used in early steam engines.

Other designs have to be applied to flywheels made up by fibre materials, where the tensile strength is only high in one direction. *Figure 6.13* shows such a design named a 'superflywheel'.

In order to avoid the gyro effect, which is proportional to the first power of the angular velocity, systems with flywheels turning in opposite directions have been proposed.

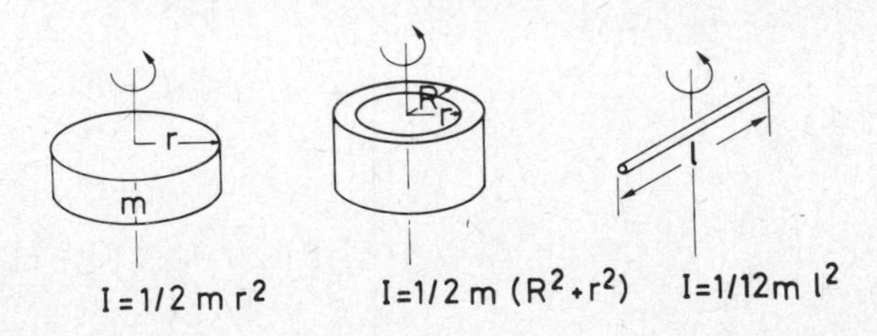

Figure 6.12
Moment of inertia for disc, cylinder and bar shaped flywheel

Figure 6.13
Proposed 'superflywheel'

A variety of flywheel systems has been tested on laboratory scale in recent years. All high angular velocity and advanced materials systems investigated are still in the early prototype stage. The ordinary steel discs flywheel, however, has been in operation in traction systems for several years. The major drawback is the low energy density (10–20 Wh/kg). Such traditional steel flywheels as the only energy store for traction purposes, therefore, have been dismissed as a possibility. As high power units in hybrid systems, on the other hand, the prospects for flywheels look good.

Further reading

Glendenning, I., 'Compressed Air Storage', IEE Proc. Int. Conf. Future Energy Concepts, London (1979)

Post, R. F., and Post, S. F., 'Flywheels', *Sci. Amer.*, 229, Vol. 6, December 1973

Rockwell International, 'Economic and Technical Feasibility Study for Energy Storage Flywheel', Final Report ERDA 76–65, UC–94B, US Govt. Printing Office, Washington DC (1975)

7 Electrical and Magnetic Storage

Electric fields

Energy can be stored in the electric field of a capacitor by virtue of the potential energy of the charges. In the act of creating an electric field, for instance between the plates of a charged capacitor, energy has to be stored. As an aid to understanding, we will adopt the point of view that the energy is actually stored in the space where the electrical field exists.

An electric capacitor is a device which can take up charges whereby an electric field is established and hence energy is stored. The capacitance C of a capacitor is defined by the amount of charge q it can take up and store per unit of voltage

$$C = \frac{q}{V_C} \tag{7.1}$$

where

C = capacitance
q = charge
V_C = the voltage of the capacitor

At a given voltage V_C the energy contained in the electric field is proportional to the capacitance, which again is dependent upon the medium in which the field exists. Some materials called dielectrics can be polarised and therefore the total charge stored in a capacitor with such materials will increase, and so will the stored energy. Let us take as an example a plate capacitor with plate area A and distance d between the plates as shown in *Figure 7.1*.

$$C = \frac{\epsilon A}{d} \tag{7.2}$$

The properties of the medium are described by the material constant ϵ called the permittivity. The unit of ϵ is farad/metre since both A and d have the metre as the basic unit. The electric field E is homogeneous within the space between the plates only when the plates are close together. In order to keep the charges at the plates apart and thereby to maintain the field, the medium

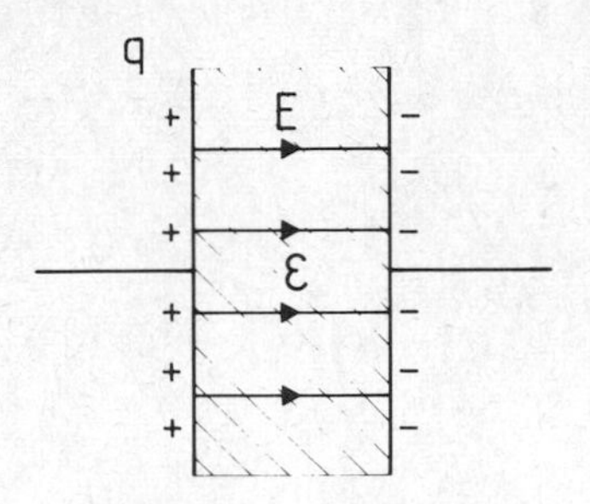

Figure 7.1
Parallel plate capacitor

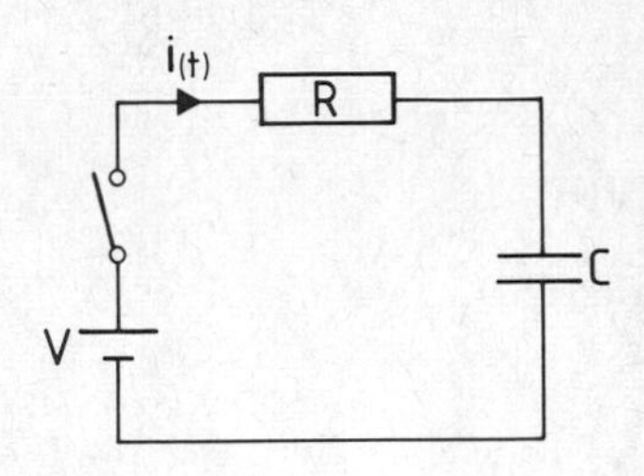

Figure 7.2
Series *RC* circuit

must have a low electronic conductivity. That means that, for the purpose of energy storage, we are looking for insulating materials with high permittivity.

When a simple circuit consisting of a resistor and a capacitor is connected to a constant voltage source some of the electric energy from the power source will end up in the field of the capacitor as electrostatic energy. What happens in *Figure 7.2* when the switch is closed can be described by Kirchhoff's circuit equation:

$$V - i_{(t)}R - V_C = 0 \tag{7.3}$$

and since $V_C = q/C$

$$V - i_{(t)}R - q/C = 0 \tag{7.4}$$

$$V - i_{(t)}R - i\,\mathrm{d}t/C = 0$$

by differentiation and rearranging

$$-R\frac{\mathrm{d}i_{(t)}}{\mathrm{d}t} - \frac{i_{(t)}}{C} = 0$$

$$\int\frac{\mathrm{d}i_{(t)}}{i_{(t)}} = -\int\frac{\mathrm{d}t}{RC}$$

∴

$$i_{(t)} = I_0\exp(-t/RC) \tag{7.5}$$

where

$i_{(t)}$ = circuit current at a given time t
I_0 = circuit current at time equal to zero
t = time
R = total circuit resistance

The graph corresponding to Equation (7.5) is shown in *Figure 7.3*. The current at $t = 0$ is only limited by the resistance R.

$$I_0 = \frac{V}{R} \tag{7.6}$$

From *Figure 7.3* we can find the time constant τ of the process. The time constant is defined as the time taken to end the process if the rate of change was constant. Hence the tangent to the current

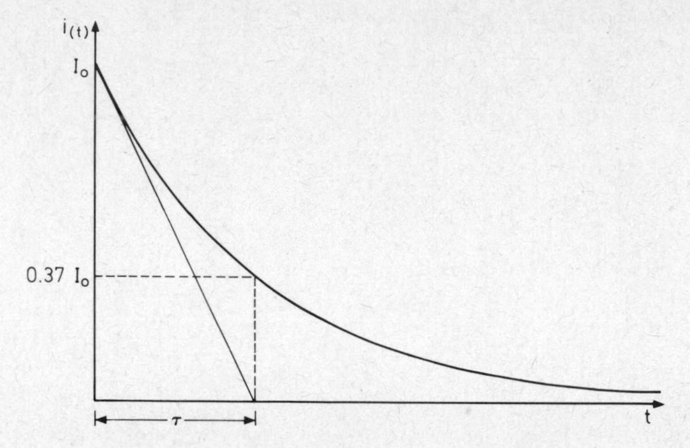

Figure 7.3
Transient response of a series RC circuit

curve at $t = 0$ will cut off the number of seconds τ on the time axis. Mathematically we can calculate τ by differentiating Equation (7.5), taking into account that the value of $i_{(t)}$ at $t = 0$ is as given in Equation (7.5). Hence

$$\tau = R\,C \tag{7.7}$$

From Equation (7.5) we can calculate the current at $t = \tau$, the result being $0.37\ I_0$. This is an important feature when designing electronic circuits where capacitors are part of a time relay function. The time it takes to charge a capacitor is, as a practical rule, equal to 5τ since by that time almost the whole charging process has been completed. The current as a function of time for discharging is exactly the same as for charging and so is τ too.

Let us now consider the main purpose of this chapter, namely to calculate the amount of stored energy. We will here return to Equation (7.4).

$$V - i_{(t)}R - q/C = 0$$

This equation, which is a voltage equation, can be transformed to an energy equation by multiplying each term by charge, since voltage is defined as work or energy W per charge unit.

$$V = \frac{\mathrm{d}W}{\mathrm{d}q} \quad \text{or} \quad \mathrm{d}W = V\,\mathrm{d}q \tag{7.8}$$

The energy equation therefore will be

$$V\,\mathrm{d}q - i_{(t)}R\,\mathrm{d}q - (1/C)\,q\,\mathrm{d}q = 0$$

or

$$\mathrm{d}W_\mathrm{T} = \mathrm{d}W_\mathrm{R} + \mathrm{d}W_\mathrm{C} \tag{7.9}$$

Equation (7.9) tells us that of the total amount of energy $\mathrm{d}W_\mathrm{T}$ drawn from the power supply some is wasted as heat in the resistor $\mathrm{d}W_\mathrm{R}$ and some is stored in the capacitor $\mathrm{d}W_\mathrm{C}$. The latter term is given by

$$\mathrm{d}W_\mathrm{C} = \frac{1}{C}\,q\,\mathrm{d}q \tag{7.10}$$

The solution of which is

$$W_C = \tfrac{1}{2}\frac{q^2}{C} \tag{7.11}$$

By substituting from Equation (7.1) we can get two other expressions for the energy stored in a capacitor, namely

$$W_C = \tfrac{1}{2} C V^2 \tag{7.12}$$

and

$$W_C = \tfrac{1}{2} Q V \tag{7.13}$$

We can now calculate the volume energy density w of the homogeneous field in the parallel-plate capacitor. Here the relation between the electrostatic field and the voltage is

$$E = \frac{V}{d} \tag{7.14}$$

The volume energy density is by definition

$$w = \frac{\mathrm{d}W}{\mathrm{d(Vol)}} \tag{7.15}$$

and for w = constant (which is the case in a homogeneous field)

$$w = \frac{W}{\mathrm{Vol}} = \tfrac{1}{2}\frac{C V^2}{d A}$$

By substituting C from Equation (7.1) and V from Equation (7.14) we get

$$w = \tfrac{1}{2}\frac{\epsilon A E^2 d^2}{d\, d A} = \tfrac{1}{2}\epsilon E^2 \tag{7.16}$$

The amount of stored energy is as mentioned earlier proportional to the material's constant ϵ or in other words the materials ability to polarise when placed in the electric field between the plates of the capacitor. Most insulators have a relative permittivity ϵ_r from 1–10 compared to that of vacuum or dry air ϵ_0. The relation being

$$\epsilon = \epsilon_0\, \epsilon_r \tag{7.17}$$

All materials which consist of dipoles can polarise and they have a ϵ_r value, also called the dielectric constant, greater than one. Teflon, for instance, with a ϵ_r value about two, will double the energy content when inserted between the plates of a capacitor. Some compounds, such as titanates, have ϵ_r values as high as 10 000–15 000, and they are therefore often used as dielectric materials in capacitors especially where small size and large capacitance are required.

Certain types of dielectric materials show hysteresis in the polarisation produced by external fields. Hysteresis means that the polarisation depends not only on the test field presently applied, but also on the field applied earlier. Such materials are called ferroelectric by analogue with the somewhat similar magnetic effect in ferromagnetic materials (see next Section on magnetic fields). One of the best-known ferroelectric materials is barium titanate ($BaTiO_3$). In this material, as in other ferroelectrics, the local fields set up by small ion displacements produce forces that are greater than the elastic restoring forces within the crystal. The equilibrium position of the ions is such as to give a net polarisation in the crystal and that points to an application for information storage. Ferroelectric materials will in the future presumably replace some of the magnetic materials in computer memory systems. Information storage and memory systems, however, are not the main concern.

Returning to the key energy equations in order to answer the question: how much energy can be stored in electric fields? We will do this by putting numbers from practical life into the equations and calculating how many joules are stored in a device.

The result of our calculations shows that ordinary capacitors can store only very limited amounts of energy, because of their small capacitance. The farad is a rather large unit and in practice the size of a capacitor is labelled in units of picofarad (1 pF = 10^{-12} F) or microfarad (1 μF = 10^{-6} F). Even the whole earth as a spherical capacitor represents a capacitance in the order of magnitude much less than 1 μF. A well known exercise for undergraduate electrical engineering students is to calculate how much energy the radial field of the earth contains if the earth was charged to, say, thousands of volts relative to the far outer space. Every student is surprised by the result of the calculation being in the order of magnitude of one joule.

For our calculation we will choose a 1 μF capacitor and a voltage of 100 V.

$$W_C = \tfrac{1}{2} C V^2 = \tfrac{1}{2} 10^{-6} \times 10^4 = 5 \times 10^{-3} \text{ J}$$

Five millijoule certainly is very little and we may ask: how about the energy density of electric fields? Is that also very small? As an example let us take a 10^7 V/m field in a good insulator with $\epsilon = 10^{-11}$ F/m.

$$w = \tfrac{1}{2} \epsilon E^2 = \tfrac{1}{2} 10^{-11} \times 10^{14} = 500 \text{ J/m}^3$$

500 J/m^3 is equivalent to 0.14 Wh/m^3 and this, of course, is a low value compared to other electricity storage devices, e.g. electric batteries. But the power density can be extremely high when a capacitor is short circuited, much higher than any battery. This means that capacitors find important applications as storage and power supply when very high power is required. One should not exclude capacitors as future energy storage units for such applications. The development of solid ionic conductors in recent years presents the prospect for capacitances in the range of one farad in a volume of only one cubic centimetre and that is many orders of magnitude more than even the best dielectric materials. Such components, where electrochemical processes are also involved, can very well be of future importance, e.g. in vehicle regenerative

braking systems, where the energy requirement is moderate and the power requirement is essential.

Magnetic fields

Magnetic fields, too, can be used for energy storage. When an electromagnet is connected to a constant voltage source, the energy flow into the magnetic system varies with time. The electric current is zero at $t = 0$ and it stabilises at I_{st} when the magnetic field has been built up. The energy supplied from the external source divides as we shall see into two parts, one of which is the energy content of the magnetic field. It takes energy to build this field up and that energy can be released again as an electric current in an outer circuit.

We have seen that the electrostatic energy in a capacitor depends on the applied voltage. The energy content in an electromagnet is determined by the current I_{st} through the N turns of the coil of the magnet. The product NI_{st} is called the magnetomotive force and this quantity is usually denoted by the Greek letter θ. We have now an analogue to the electric circuit where the electromotive force is equal to the voltage drops IR throughout the length l of the closed electrical circuit. The analogue named Amperes rule is

$$\theta = \Sigma H l = H_1 l_1 + H_2 l_2 + \ldots H_n H_n \tag{7.18}$$

Here H is the magnetic field and l is the length of the field in which H exists. Another analogue is Ohm's law for a magnetic circuit as shown in *Figure 7.4*.

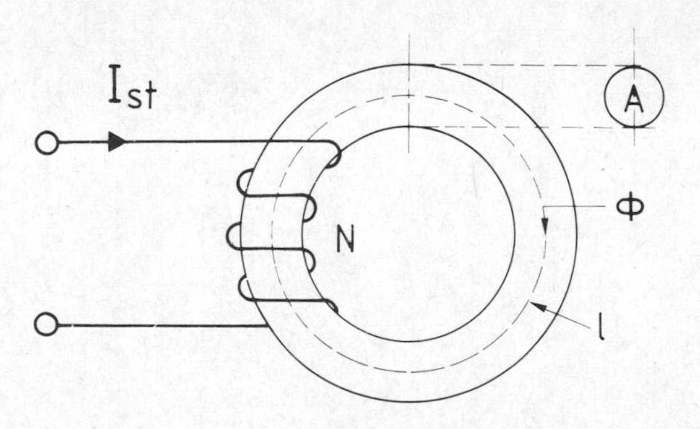

Figure 7.4
The magnetic circuit

$$\theta = \Phi R_{magn} \tag{7.19}$$

$$\Phi = B_n dA = B A \quad \text{for constant } B \tag{7.20}$$

$$R_{magn} = \frac{l}{\mu A} = \frac{N^2}{L} \tag{7.21}$$

where

θ	=	magnetomotive force
Φ	=	magnetic flux
B	=	induction
R_{magn}	=	magnetic resistance
l	=	length of the magnetic field
A	=	area of the magnetic field
N	=	number of turns of the coil
μ	=	permeability
L	=	self-inductance

When earlier we described the properties of the electric field we took as an example the simple parallel-plate capacitor. Here we also simplify a little by making a few assumptions. Firstly in Equation (7.20) we assume a magnetic circuit where B has the same value throughout the area A. This assumption holds for all magnetic circuits with limited area compared to length. Secondly we assume a linear relationship between H and B or, in other words, μ is taken as a material constant, the relation being:

$$B = \mu H \tag{7.22}$$

This simplified linear relationship is often used in magnetic calculation, but in real ferromagnetic materials, $B = f(H)$ does not have constant slope, and we may consider μ as the linearised value of the slope which is shown in the magnetisation curve in *Figure 7.5.* Fortunately this assumption is good enough when dealing with calculations of energy content.

The magnetisation curve, or rather the hysteresis loop shown in *Figure 7.5*, suggests application in a memory device. If we follow the heavy line at the right hand of the figure we see that by decreasing the electric current from I_{st} back to zero there is still some induction B left. The value reflects the prehistory of the magnetisation and so we could name it the memory m of the magnetic material. The material 'remembers' whether it has been

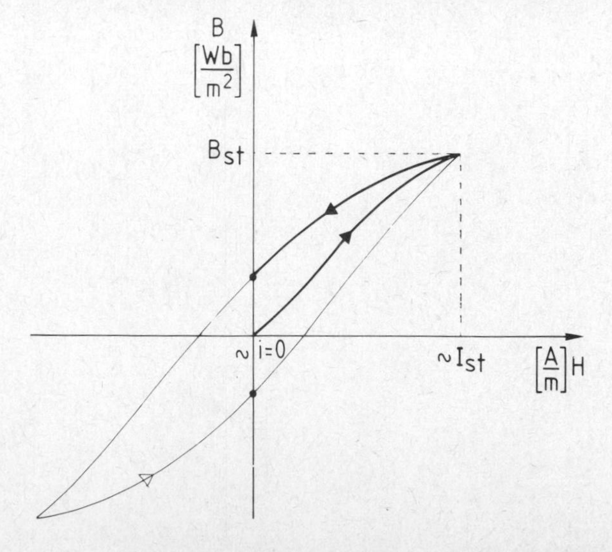

Figure 7.5
Hysteresis loop characteristic of a ferromagnetic specimen

magnetised to the right hand side or the left hand side and correspondingly it ends up at either m_1 or at m_2. This information storage ability of ferromagnetic materials is widely used in computer memory systems.

Again memory systems do not require much energy stored and we will therefore leave any further discussion of these systems and return to the question: what determines the energy content of an electromagnet? The answer is that the higher flux Φ, the higher energy content. Let us, by rearranging the Equations (7.18)–(7.21), see what is required to obtain a high flux.

$$\Phi = \frac{\theta}{R_{magn}} = \frac{I_{st} N \mu A}{l} = G \mu I_{st} \tag{7.23}$$

By a given design, indicated by the geometry factor G, we obtain high flux when the stationary electrical current I_{st} and the material's properties given by μ have high values. I_{st} for a given voltage V is limited only by the electrical resistance R of the coil

$$I_{st} = \frac{V}{R} \tag{7.24}$$

Therefore, in order to get a high flux value, low values of R and high values of μ are necessary. Modern ferromagnetic alloys exhibit μ values that are several orders of magnitudes higher than the permeability of the vacuum μ.

$$\mu = \mu_0 \mu_r \tag{7.25}$$

Hence optimisation efforts deal with both superconducting coils, i.e. coils with very low electrical resistance R, and with the development of magnetic materials which are highly permeable to magnetic fields.

We will now look a little closer at the quantitative parameters of the transient storage process and the expression of the stored energy. We do that by closing the switch in *Figure 7.6* where R is the total resistance of the electrical circuit and L the self-inductance. V is the power supply – a constant voltage source – and $i_{(t)}$ is the electrical current at a given time t.

$$V - R\, i_{(t)} + e_s = 0 \tag{7.26}$$

where e_s is the Faraday induced emf

$$e_s = -L \frac{di}{dt} \tag{7.27}$$

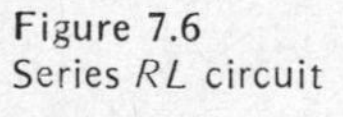
Figure 7.6
Series RL circuit

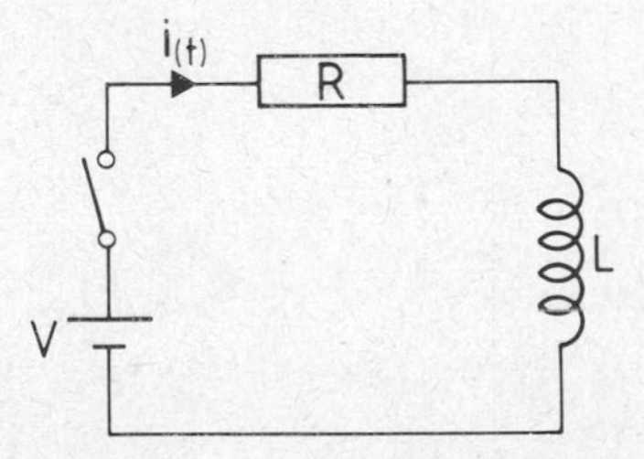

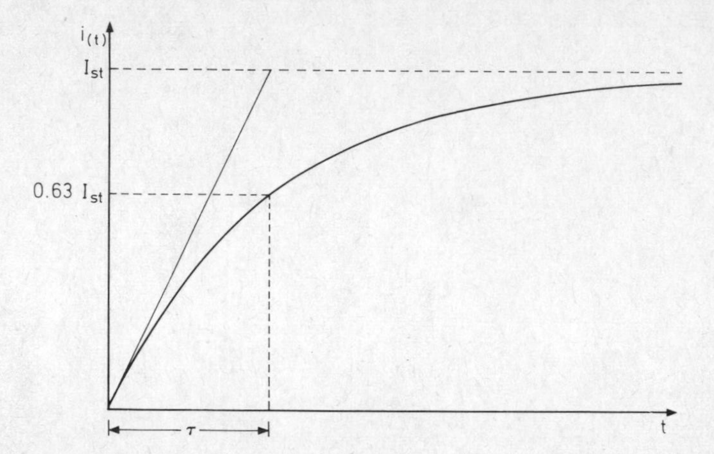

Figure 7.7
Time variation of current in a *RL* circuit

With this expression for e_s and the constraint $i_{(t)} = 0$ at $t = 0$ in mind, Equation (7.26) can be solved, the result being as shown in *Figure 7.7*

$$i_{(t)} = I_{st} \left\{1 - \exp\left(-R_t/L\right)\right\} \tag{7.28}$$

The time constant of this process is

$$\tau = L/R \tag{7.29}$$

By multiplying each term of the voltage Equation (7.26) by $\mathrm{d}q$ or $i\mathrm{d}t$ we get the energy Equation (7.30)

$$V i_{(t)}\,\mathrm{d}t - i_{(t)}{}^2 R\,\mathrm{d}t - L\,\frac{\mathrm{d}i_{(t)}}{\mathrm{d}t}\,i_{(t)}\,\mathrm{d}t = 0$$

or

$$\mathrm{d}W_\mathrm{T} = \mathrm{d}W_\mathrm{R} + \mathrm{d}W_\mathrm{M} \tag{7.30}$$

The energy stored in the magnetic field is $\mathrm{d}W_\mathrm{M}$

$$\mathrm{d}W_\mathrm{M} = L\,i_{(t)}\,\mathrm{d}i_{(t)} \tag{7.31}$$

and as L is a constant it follows that

$$W_\mathrm{M} = L\int_0^{I_{st}} i_{(t)}\,\mathrm{d}i_{(t)} = ½\,L\,I_{st}{}^2 \tag{7.32}$$

If we, in Equation (7.26), use another expression for the Faraday induced emf

$$e_s = -N\,\frac{\mathrm{d}\Phi}{\mathrm{d}t} \tag{7.33}$$

then the magnetic energy W_M is

$$W_\mathrm{M} = \int_0^{\Phi} N\,i_{(t)}\,\mathrm{d}\Phi = \int_0^{B} l\,A\,H\,\mathrm{d}B = \mathrm{Vol}\int_0^{B} H\,\mathrm{d}B \tag{7.34}$$

Assuming a linear relationship μ between H and B, the solution of this integral is

$$W_M = \text{Vol}\ \tfrac{1}{2}\ \mu H^2 \tag{7.35}$$

or

$$W_M = \text{Vol}\ \tfrac{1}{2}\ H B \tag{7.36}$$

$$W_M = \text{Vol}\ \tfrac{1}{2}\ \frac{B^2}{\mu} \tag{7.37}$$

Since the volume (Vol) of the magnetic field is an explicit part of the Equation (7.35)–(7.37), the volume energy density is obtained from these equations

$$w = \tfrac{1}{2}\mu H^2 = \tfrac{1}{2} H B = \tfrac{1}{2}\frac{B^2}{\mu} \tag{7.38}$$

Choosing $B = 1.0\ \text{Wb/m}^2$ and $\mu = 7 \times 10^{-4}$ H/m we get

$$w = \tfrac{1}{2}\frac{B^2}{\mu} = \tfrac{1}{2}\frac{1.0^2}{7 \times 10^{-4}} = 716\ \text{J/m}^3$$

Clearly this magnetic density is higher than the one calculated for the electrostatic field, but it is not too impressive to be able to store only a few hundred joule in a cubic metre. Accordingly the total amount of stored energy in an electromagnetic device is rather limited. For example in the field system of a large ac-motor with a self-inductance L as high as 1 H/m and an electric current of 10 A, W_M is as follows

$$W_M = \tfrac{1}{2} L\, I_{st}^2 = \tfrac{1}{2}\, 1 \times 10^2 = 50\ \text{J}$$

Hence the conclusion is bound to be that ordinary electromagnets only store a limited amount of energy. So we have to look for extraordinary electromagnets in order to get useful energy values. We will consider the so-called superconducting magnet, bearing in mind Equation (7.23) told us that we are looking for a high I_{st}, such as is found in superconducting coils. Some conductors when cooled close to zero degrees Kelvin exhibit zero electrical resistance R and since R is the only steady state limiting factor on I_{st}, very high current values can be obtained almost without loss of power.

Superconducting coils

The ability of a superconductor is to carry high currents in the presence of high magnetic fields with zero resistance to the steady flow of electrical current points towards applications involving energy and power. The high current density allows a device to be considerably more compact when compared with conventional devices designed for the same application.

Present superconducting materials, such as intermetallic compounds and alloys, have critical temperatures ranging from

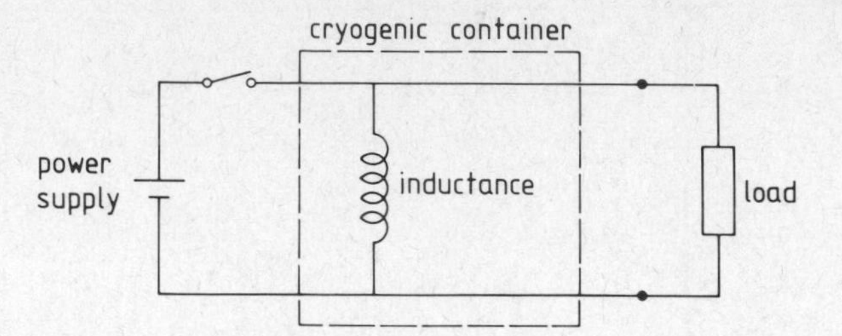

Figure 7.8
Simplified superconducting energy storage circuit

10–20 K and the penalty paid for the zero resistance and compact character is the need for operation at liquid-helium temperature with the associated problem of using vacuum-insulated cryogenic containers. In general, this places a lower limit on the size of the device below which the superconducting device is not practical.

A superconducting coil can be connected to a constant dc power supply as shown in *Figure 7.8*. As the current of the coil (which is a pure inductance) increases, the magnetic field also increases and all electrical energy is stored in the magnetic field. Once I_{st} is reached, the voltage across the coil terminals is reduced to zero. At this stage the system is fully charged, and the energy can be stored as long as desired. In contrast, a conventional coil made of copper windings which exhibit electrical resistance would require continuous power input to keep the current flowing.

The applications of superconducting coils for energy storage are the following:

generation of high power pulses of electrical energy (millisecond range)

part of proposed fusion reactors instead of capacitor banks (millisecond discharge times)

storing of energy between pulses of a high energy particle accelerator (charge and discharge in a few seconds)

part of a flywheel discharge system instead of conventional electrical conversion systems (charge and discharge in the range of seconds or minutes)

utility peak load shaving (charge and discharge in hours)

long term storage (large volume and high magnetic field)

An important capability of superconducting coils is that they can store energy at a lower power level for later discharge at a higher power level. Few of the above mentioned applications are now in full use (see Chapter 8), but their future potential is excellent.

Further reading

Stekly, Z. J. J. and Thome, R. J., 'Large-Scale Applications of Superconducting Coils', *Proc. IEEE*, Vol. 61, No 1 (1973), Pp 85–95

Voinov, M., 'Various Utilisations of Solid Electrolytes, Energy Storage Devices', *Electrode Processes in Solid State Ionics, Theory and Application to Energy Conversion and Storage*, Edited by Kleitz, M. and Dupuy, J., Reidel (1976), Pp 431–464

8 Some Existing Storage Systems and Cost Estimates

As mentioned in the introduction a filled oil tank is a convenient and cheap way of storing energy compared to the alternatives which have been discussed in this book. This statement is well illustrated by the overview of energy density and capital cost made by Dr. P. R. Smith of Rutherford Laboratory, UK (see *Figure 8.1*).

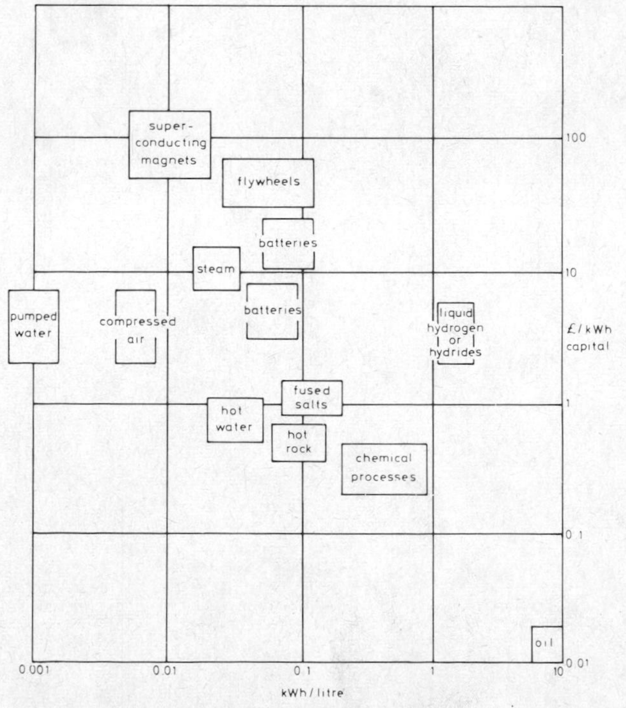

Figure 8.1
Energy density and capital cost of energy storage systems. Note that some of the data for the advanced systems is estimated

Table 8.1 Expected technical and cost characteristics of selected energy storage systems (Mallard, S.A. *et al.*, *Proc. Am. Power Conf.*, Vol. 38 (1976) p. 1200)

	Near term					Intermediate term			Long term
			Thermal						
Characteristics	*Hydro pumped storage*	*Compressed air*	*Steam*	*Oil*	*Lead acid batteries*	*Advanced batteries*	*Flywheel*	*Hydrogen storage*	*Super-conducting magnetic*
Commercial availability	Present	Present	Before 1985	Before 1985	Before 1985	1985–2000	1985–2000	1985–2000	Post 2000
Economic plant size (MWh or MW)	200–2000 MW	200–2000 MW	50–200 MW	50–200 MW	20–50 MWh	20–50 MWh	10–50 MWh	20–50 MW	Greater than 10 000 MWh
Power related costs[(a)] ($/kW)	90–160	100–210	150–250	150–250	70–80	60–70	65–75	500–860	50–60
Storage related costs[(a)] ($/kWh)	2–12	4–30	30–70	10–15	65–110	20–60	100–300	6–15	30–140[(c)]
Expected life (years)	50	20–25	25–30	25–30	5–10	10–20	20–25	10–25	20–30
Efficiency[(d)] %	70–75	(e)	65–75	65–75	60–75	70–80	70–85	40–50	70–85
Construction lead time (years)	8–12	3–12	5–12[(f)]	5–12[(f)]	2–3	2–3	2–3	2–3	8–12

(a) Constant 1975 dollars, does not include cost of money during construction (b) Could be considerably higher (c) These numbers are very preliminary (d) Electric energy out to electric energy in, in percent (e) Heat rate of 4200–5500 Btu/kWh and compressed air pumping requirements from 0.58–0.80 kWh (IN)/kWh (OUT) (f) Long lead time includes construction of main power plant

In this chapter some existing alternatives will be described and actual cost or cost estimates will be included where such information is available. Many of the alternatives are, however, still at the development stage and therefore real cost figures are difficult to obtain. *Table 8.1* shows some estimates.

Heat storage

Autonomous solar house, Cambridge (England) with hot water store

Project: A. Pike, J. Thring.
Calculations: G. Smith, J. Littler, C. Freeman, R. Thomas.

This solar house project, carried out by a working group of the University of Cambridge, is the result of three years of research work between 1971 and 1974. The result of the study was, that a fully autonomous house with comfort comparable to that which we know today, can be achieved and is also economical.

In this project all locally available sources of energy are utilised. Solar collectors produce heat and distilled drinking water, a wind generator provides electricity for the kitchen, lighting and heat pumps, and the sewage disposal system produces methane. An indoor garden behind the south face produces oxygen and food, and can also be used for most of the year as an 'open space'.

Figure 8.2
Autonomous solar house in Cambridge (England 52° 12′N.)
1. Radiation
2. Solar collectors (40 m^2)
3. Wind generator
4. Hot water store (about 10 m^3)
5. Interior garden
6. Living space (total 111 m^3)
7. Bedroom
8. Insulating wall to the North

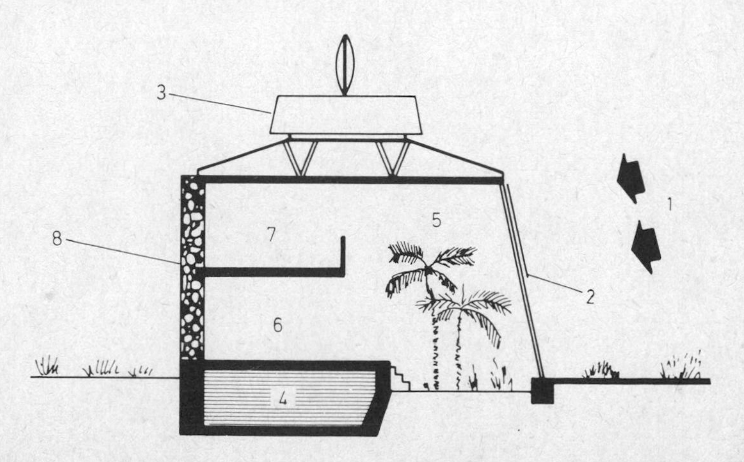

The house has a gross floor area of about 111 m^2. The living space is on two floors and the garden extends up to the roof (*Figure 8.2*).

The room heating requires 61 to 75 % of all the energy and uses low grade heat – 50 to 100 W/m^2 at temperatures between 15 ° and 30 °C. The solar radiation also amounts to a similar value (100 to 150 W/m^2). The solar collectors are of surface type and have an area of 40 m^2.

The heat transfer and storage medium is water. The heat store is in the cellar, and calculations have given an optimum size of 10 m^3. In November 1974, a one tenth scale model of the house was produced with which computer simulations are possible to enable optimum dimensions for the units to be determined.

'Solar One House', Delaware (USA) with phase change material as storage medium

Project: K. Böer, M. Telkes, K. O'Connor.
Built: 1973

The solar house of the 'Institute of Energy Conversion' University of Delaware, is claimed to be the first in the world in which the solar radiation, as well as being converted into heat, is also converted directly into electrical power. The building was financed collectively by eight research institutes and electricity supply undertakings. The cost of the building, excluding solar cells, was 130 000 dollars (*Figure 8.3*).

The total living space of the house is 132 m^2; the living rooms are on two levels. The air type solar collectors are mounted on the 45 ° inclined roof and on the south face. These solar collectors have a total area of 82 m^2, with a double Plexiglass covering. Parts of the solar collectors are combined with solar cells (cadmium sulphide – copper sulphide cells) which have a maximum output of 19 mA/cm^2 at a voltage of 0.37 V. The efficiency of the direct conversion is 6–7 %. The life of the solar cells is estimated at ten years.

The total conversion factor of the collector is 50 %, of which 45 % is converted into heat and 5 % into electrical power. The house receives 80 % of its energy requirements from the sun and the remaining 20 % from electrical power. The heat storage system works on a chemical principle with three different salt solutions which have a low melting point between + 24 °C and + 49 °C. The heat transfer from collector to store, and transfer from store to the living room is by air moved by fans. A heat pump is also associated with the system. The lead/acid accumulators of the electrical storage system have a capacity of about 20 kWh. The solar installation also provides cooling in summer. According to the calculations of Aaron and Isakoff it should soon be possible, through mass production, to be able to make combined solar cells/collectors for 10/m^2 dollars. The chemical heat storage of a similar villa should not cost more than 900 dollars.

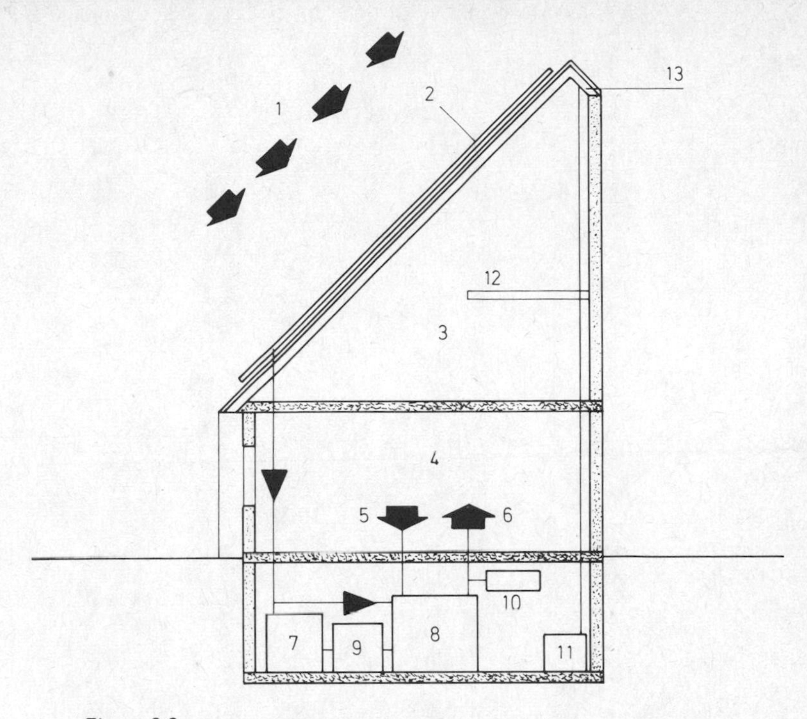

Figure 8.3
'Solar One House' in Delaware (U.S.A. 39° 35'N.)

1. Radiation
2. Solar collectors (total 82 m^2)
3. Thermal buffer zone
4. Living space (total 132 m^3)
5. Return air
6. Warm air to living space
7. Chemical auxiliary store ($Na_2S_2O_3 5H_2O$)
8. Chemical main store ($Na_2S_2O_3 5H_2O$) 3600 kg. 49°C, 235 kWh
9. Heat pump
10. Electrical auxiliary heating
11. Accumulator 180 Ah
12. Installation support
13. Connection to electrical supply

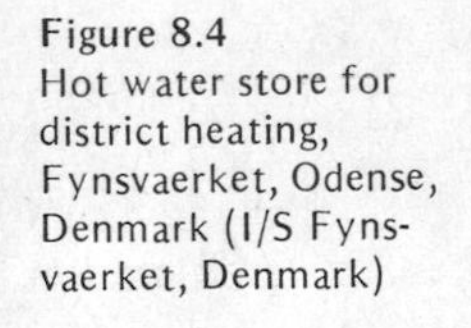

Figure 8.4
Hot water store for district heating, Fynsvaerket, Odense, Denmark (I/S Fynsvaerket, Denmark)

District heating with large heat store

About 80 % of all heating in the town of Odense in Denmark comes from the 'waste heat' of the nearby power station Fynsvaerket as district heating. The power station is a so called combined heat and power station.

Early in 1978, a 12 000 cubic metre highly insulated hot water tank was put into operation – enough to give a 2 hour heat supply (see *Figure 8.4*). Before the installation of this heat store for the district heating system, the heat supply had to be shut down in short periods when demand for electricity was high and that caused temperature variations for the heat customers.

The heat storage tank added an extra capacity of 40 MW (electricity) and the cost being only £1 million makes it a good investment (25 £/kW). The yearly maintenance costs are estimated to 2.5 k£.

Underground storage of oil and methane (natural gas)

Underground bulk-storage of propane and butane was established 30 years ago in Texas, USA. The first underground gas store was established in 1971 near Kiel in Western Germany. Today there are 14 cavern stores for oil and natural gas, in northern Germany, as shown in *Figure 8.5*.

Liquid hydrogen

Energy storage in the form of liquefied hydrogen is already a routine practice in the space industry. The largest single liquid hydrogen tank is shown in *Figure 8.6*. This vacuum insulated cryogenic tank at the John F. Kennedy Space Center, USA, contains 900 000 gallons of liquid hydrogen, which has been used for fuelling the Apollo rockets. In terms of energy the content equals 11 000 MWh of electricity.

The possibility of using liquid hydrogen storage in a way analogous to liquid natural gas storage seems perfectly feasible, although sufficiently large storage tanks for seasonal storage have not been constructed yet. The tank shown in *Figure 8.6* contains less than 5 % of the energy stored in a typical liquid natural gas tank (Philadelphia Gas and Electric Co.) but is 75 % of the capacity of the largest pumped hydroelectric storage plant at Ludington, Michigan, USA.

The cost for liquid hydrogen storage will be greater than LNG because of the lower boiling temperature of hydrogen. The production cost of bulk hydrogen has been estimated to 3–5 times as much as that of oil based fuels, but as yet no commercially viable cost estimates for storage are available.

Figure 8.5
Underground stores of oil and natural gas in Northern Germany. The largest gas store, No. 14 at Epe, contains 380 million cubic metres of gas (Risø National Laboratory, Denmark)

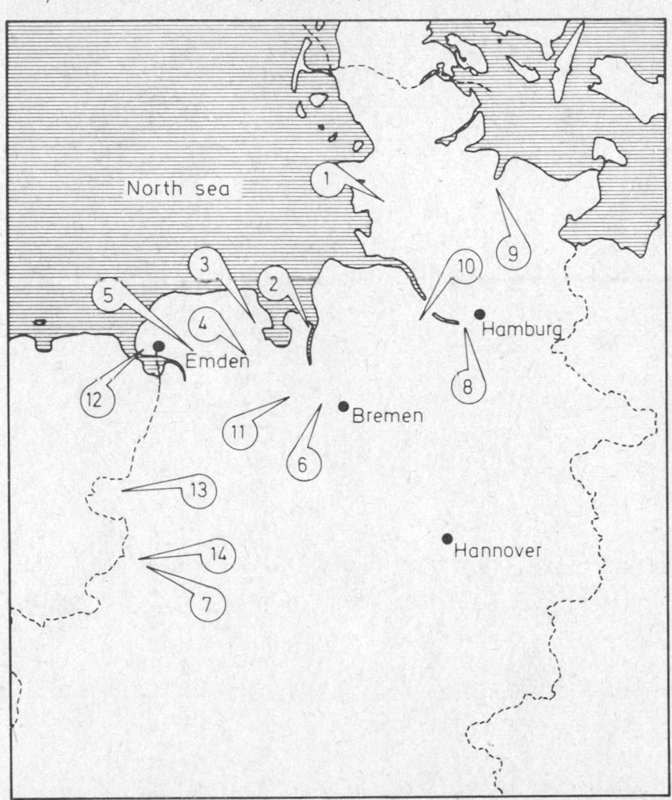

Figure 8.6
Liquid hydrogen tank at the John F. Kennedy Space Center, USA

Table 8.2 Characteristics of selected batteries (likely achievable performance) (McGeehin, P., and Jensen, J., 'Batteries for Energy Storage in Transport and Stationary Applications', *Proc. Int. Conf. on Future Energy Concepts,* London, IEE Publ. 171 (1979) Pp. 190–194

		Lead–acid	*Nickel–cadmium*	*Nickel–iron*	*Nickel–zinc*	*Iron–air*	*Zinc–air*	*Zinc–chlorine*	*Lithium–TiS_2*	*Sodium–sulphur*	*Lithium–sulphur*
Electrolyte		H_2SO_4	KOH	KOH	KOH	KOH	KOH	$ZnCl_2$	various organic	β-Al_2O_3	LiCl/LiI/KI
Voltage (V)	Open circuit	2.05	1.35	1.37	1.71	1.27	1.65	2.12	2–3	2.1–1.8	1.9–1.4
	Discharge at 2 h rate	1.9	1.2	1.2	1.6	0.7	1.2	1.85	1.9	1.7–1.4	1.3–1.0
Energy efficiency (charge-discharge) %		75	70	<60	75	40	55	65	–	70–75	75
Specific energy (Wh/kg)	1 h rate	24	28	40	70	50	80	120	–	120	~140
	5 h rate	40	30	55	75	80	100	150	>100	140	
Energy density (Wh/dm^3)	1 h rate	70	60	100	140	80	80	180	330	170	
Specific power (W/kg)	Peak	120	300	440	400	60	100	280	110	240	200
	Sustained	25	140	220	200	50	–	–	–	120	140
Life (to 80 % discharge) cycles		500	2000	2000	350	200	100	~100	~250	2000	200
Recharge time (h)		5–8	4–7	4–7	3–6	4–5	5–8	5	5	7–8	5
Operating temperature (°C)		–20 +50	–30 +50	10 50	–30 +40	0 +50	0 +60	0	<75	300–400	430–500
Cost	Capital (£/kWh)	25	250	65	30	23	75	18	24	30	30
	Running (£/kWh/cycle)	0.05	0.2	0.03	0.1	0.11	0.7	0.3	0.1	0.02	0.15
		Existing ← (Lead–acid to Nickel–iron) →			Under development ← (Nickel–zinc to Lithium–sulphur) →						
		Ambient temperature ← (Lead–acid to Lithium–TiS_2) →								High temperature ← (Sodium–sulphur, Lithium–sulphur) →	

Note: These figures should be regarded only as guidelines, since they differ with battery design and duty. Some are projections from cell performance only.

Table 8.3 Target capital costs for utility batteries (McGeehin, P., and Jensen, J., 1979)

Daily discharge period	*Battery life 10 years*		*Battery life 20 years*	
	Target cost ($/kW)	*Target cost* ($/kWh)	*Target cost* ($/kW)	*Target cost* ($/kWh)
2	50–100	25–50	80–160	40–80
5	75–175	15–35	125–300	25–60
10	100–300	10–30	200–500	20–50

Note: The range for each pair of figures reflects wide variation in possible ratios of peak to off-peak energy cost. For peak energy the crucial variables are generator capital and fuel costs. For off-peak energy, the main variable is fuel cost for the baseload generators used to charge battery storage.

Batteries

Table 8.2 shows principal data together with capital and running cost estimates. For comparison, *Table 8.3* includes the target capital cost for utility batteries.

The major market for batteries is estimated to be the electric vehicle application. Existing electric vehicles have used lead–acid batteries almost exclusively until now (see *Table 8.4*). A few of the listed vehicles are shown in *Figures 8.7–8.10.*

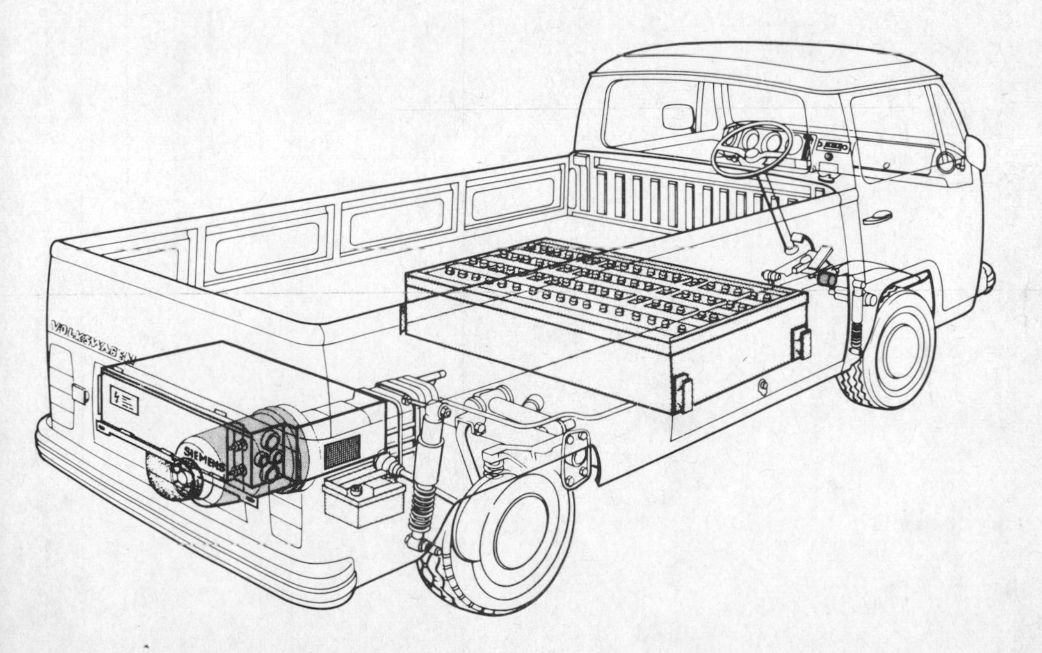

Figure 8.7
VW transporter (Volkswagen, West Germany)

Table 8.4 Current electric vehicles – Europe, USA and Japan

Vehicle name and builder	Type	Range[b] (km)	Max. speed (km/h)	Total weight (kg)	Payload (kg) or passengers	Battery		
						Weight (kg)	Capacity (kWh)	Type
Italy								
Fiat 900T	Van	55	60	1857	370	460	20	Pb–acid
	Microbus	55	60	1857	4	460	20	Pb–acid
Fiat 242	Van	85	60	3500	1000	920	37.5	Pb–acid
Fiat city car	Car	50	75	820	20 kg + 2	166	10	Ni–Zn
Vespa electrocar	Truck	50	45	1268	380 kg + 1	360	13	Pb–acid
United Kingdom								
Harbilt HSV3	Van	80	53	1640	340	770	20	Pb–acid
Lucas TP41		~160	~96	2200				Pb–acid
Midi	Bus	~160	80	9870	34			Pb–acid
Van	Van or Transporter	80+	80+	3500	6 or 700 kg	1000	31	Pb–acid
Chloride Silent Rider	Bus	64	64	~16000	50	4400	109	Pb–acid
Chloride Chrysler/NFC	Van	64	64	7500	1780	1672	58	Pb–acid
Enfield 8000	Car	40	64	1120	2 + 20 kg	310	9	Pb–acid
Germany								
MAN	Bus	40	70	16000	99	4050	101	Pb–acid
VW transporter	Van	70	70	2670	800	860	22	Pb–acid
Daimler Benz transporter	Van	60	75	4400	1450	1060	33	Pb–acid

France								
COB	Small van	40	47	760	1 + 167 kg	224	7	Pb–acid
Renault–EDF R4		55	60	1200	2 + 40 kg	330	13	Pb–acid
Renault R5		55	60–70	1252	2 + 100 kg	328	13	Pb–acid
CGE–Gregoire		60	60	1000	2 + 40 kg	300	12	Pb–acid
Sovel AS 9	Bus	50	45	9000	21	2200/2736	70/85	Pb–acid
Sovel 3T1	Bus	80	60	13450	50	4080	150	Pb–acid
USA								
American motors	Postal	50	55	1950	1.7 m^3	614	18	Pb–acid
General DJ–JE	Jeep							
General Electric	Van	48	48	1500		391	15	Pb–acid
Copper electric town car	Car	100	90	1500	2	450	12	Pb–acid
Electra	Van	100	88	1480	2 + 400 kg	440	21	Pb–acid
Japan								
Daihatsu Hijet	Van/Truck	60(a)	75	~1200	200 kg + 2			Pb–acid
Toyota Bongo	Van/Truck	50(a)	65	1570	~250 kg + 2			Pb–acid
Toyota BP30	Car	65(a)	70	1250	5			Pb–acid
UP 100E	Truck	80(a)	60	1515	350 kg + 2			Pb–acid
Nissan Cherry	Truck	70(a)	50	1425	400 kg + 2			Pb–acid
Laurel	Car	55(a)	85	1920	5			Pb–acid
Mitsubishi E30	Truck	70(a)	40	1095	100 kg + 2			Pb–acid
E12	Van	75(a)	80	915	2			Pb–acid
ME460	Bus	170(a)	60	14200	70			Pb–acid

Sources: Principally manufacturers' catalogues, Proc. the Fourth Inter. Electric Vehicle Symposium, Dusseldorf, 1976, Electric Vehicle News. (a) at constant 40 km/h (b) in urban driving conditions

Figure 8.8
Standard Vauxhall Bedford van, converted to an electric vehicle by Lucas (UK)

Figure 8.9
The FIAT mini electric car (FIAT, Italy)

Figure 8.10
The Chloride Silent Rider bus (Chloride Silent Power Ltd, UK)

Figure 8.11
Fuel cell testing. On the left is an advanced developmental low-temperature asbestos matrix fuel cell. An uprated Apollo Block II fuel cell is in the centre and on the right an Apollo fuel cell

Fuel cells

Pratt & Whitney Aircraft, Division of United Aircraft Corporation, USA, supplied the fuel cells for the Apollo programme (see *Figure 8.11*). Details of the design for the fuel cell project have not yet been fully published. However, principal parameters are given as follows:

Number of cells	31
Cell pressure	50 psia
Nominal temperature	400 °F (204 °C)
Reactant gas pressure	10 psi above cell pressure
Heat and water removal	By hydrogen circulation
Voltage	27–31 V
Power	563–1420 W
Duration	400 h
Maximum power	2295 W at 20.5 V
Wh/lb reactants	1220 (at 1420 W)
Weight	220 lb

The Apollo 8 moon flight lasted for 440 h and during that time 292 KWh of electricity were produced together with more than 100 litres of water. No exact costs are available but a conservative estimate is $100 million.

Pumped hydro

At the moment hydro-electric power is the biggest non-dissipative energy producer providing about 1.2×10^5 MW out of an estimated potential of 3×10^5 MW. In a number of small hydro-electric schemes, dams have been built in areas where the total water catchment scarcely justifies the capital cost of the equipment. For these schemes the essential economic justification lies in their reversibility.

An example of such an installation is the North of Scotland Hydro-Electric Board Station Cruachan shown in *Figure 8.12.* This station was put into operation in 1965. Cruachan is a 400 MW reversible pumped-storage development which uses energy from steam generating stations at times when the system load is low to pump water from Loch Awe to a high-level reservoir on Ben Cruachan. Pumping is carried out mostly at night and at weekends. The water stored in the high-level reservoir is then used to generate

Figure 8.12
The Cruachan Station. The dam and intake/outfall (North of Scotland Hydro-electric Board, UK)

electricity to meet daytime peak loads on the Scottish supply system.

The Cruachan power station is operated to give an annual output of some 450 million units of electricity (kWh). The operating head of the pump/turbines is 365 m, originally the highest in the world for this type of plant. This high head, in conjunction with a horizontal distance of only 1400 m between the upper and lower reservoirs, offers particularly advantageous conditions for economic pumped-storage development. Because the surface area of Loch Awe is 39 km^2 the operation of the station has little effect on the water levels. The power of the pumping installation, however, is such that if all machines were operated continuously for 24 hours to fill the upper reservoir, an operating condition which does not occur in practice, the level of Loch Awe would be lowered by 0.23 m.

The main engineering features of the works are a reservoir created in a corrie on Ben Cruachan from which two inclined shafts supply water to four reversible pump/turbines in an underground power station 396 m below the surface. A single tailrace tunnel carries the water from the station to an outfall at Loch Awe. Access to the power station is provided by a road tunnel 1 km long. A shaft rising vertically from the power hall carries the main cables to a transmission line and serves as a ventilating

Figure 8.13
Cruachan machine hall (North of Scotland Hydro-electric Board, UK)

air intake. Access to the dam is by a 5 km road along the shoulder of Ben Cruachan.

The upper reservoir is created by a massive buttress type dam containing 88 600 m^3 of concrete. The dam is 316 m long and has a maximum height of 46.6 m. The crest of the dam is 400.8 m above sea level. The reservoir has an operating range of 29 m and provides a usable storage of 10 Mm^3 which is equivalent to 8.3 million units of electricity.

The turbine/generator sets as shown in *Figure 8.13* are housed in a machine hall formed by excavating out of the solid rock a cavern nearly 91 m long and 36 m high – enough for a seven-storey building. The loading bay at the east end of the station is 36 m below the normal level of Loch Awe and the centre lines of the turbine spiral casings are a further 9.75 m below the loch.

The 134 000 hp vertical reversible Francis pump/turbines and motor/generators, developed by two different manufacturers, differ in design details. The motor/generators are rated at 100 MW as generators and as motors absorb 110 MW when pumping. Each complere machine weighs about 650 tonnes, the rotating parts alone accounting for 250 tonnes. Two sets run at 600 rev/min and two at 500 rev/min.

The station has a 15–20 % generating load factor and a 20–25 % pumping load factor. Spinning reserve amounts to 10 % in both directions. In 1977 alone about 12 000 changes of operating mode were recorded.

The instalment cost for the 400 MW capacity was £15.55 million or 43.8 £ per kW.

Compressed air storage

In 1978 the world's first air storage gas turbine plant with an output of 290 MW was put into operation at Huntorf for Nordwestdeutsche Kraftwerk AG. The plant has been built by Brown

Figure 8.14
Model of the Huntorf air storage gas turbine plant built by Brown, Boveri (Brown, Boveri et Cie, West Germany)

Boveri at a cost of 96 £/kW (1974 price). A model of the plant is shown in *Figure 8.14*.

Superconducting coils

The Baseball II superconducting magnet built for controlled thermonuclear reaction research at the Lawrence Radiation Laboratory, USA, as shown in *Figure 8.16* has a useful volume of approximately 1 m in diameter. The coils have a design field of 7.5 Wb/m^2 at the winding and the operating current density is 4000 A/cm^2. The Baseball II coil system provides a so called minimum *B* magnetic field configuration which serves to confine

Figure 8.15
Lawrence Radiation Laboratory's Baseball II superconducting magnet (Lawrence Livermore Laboratories, University of California, USA)

Figure 8.16
The liquid helium cooled superconducting BEBC magnet at CERN, Geneva (European Organization for Nuclear Research, Switzerland)

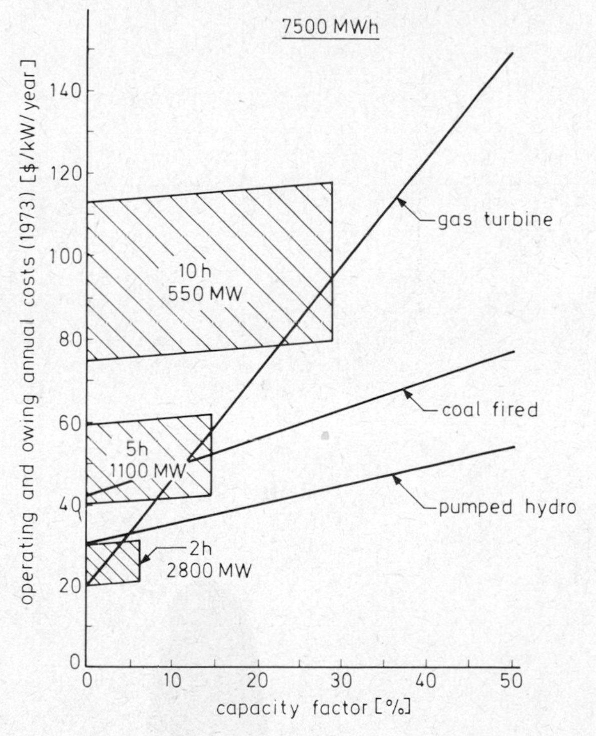

Figure 8.17
Comparison of cost of superconducting energy storage with other means of peak power shaving (Z.J.J. Stekly, Magnetic Corp. of America)

the energetic particles near the centre of the system to allow interaction.

In Europe, two large magnets have been built for Organisation Européenne pour la Recherche Nucléaire (CERN) in Geneva. The largest of these coils, the Big European Bubble Chamber (BEBC)

Magnet shown in *Figure 8.16* has an inner diameter of 4.6 m and a central field $B = 3.5$ Wb/m^2. This corresponds to a stored energy of 800 MJ or 0.25 MWh.

No superconducting magnets for utility peak power shaving have yet been built. In 1973, the Magnetic Corporation of America, Mass, USA, made a study on an underground 7.500 MHh energy storage system. *Figure 8.17* shows the estimated owing and operating cost for the superconducting system compared with gas turbine, pumped power, and coalfired peak power units.

Because the superconducting coil can be discharged either slowly or rapidly, the economics are presented for discharge times of 10.5 and 2 h, respectively. Since only the power conditioning equipment changes as the discharge time changes, the cost per unit power decreases as the discharge time decreases.

Concluding remarks

Historically there has been an approximately linear relationship between economic growth and growth in energy consumption. In view of impending oil shortages, the choice facing society is either to accept limited economic growth or to develop vigorous energy conservation and substitution methods. Such measures concern primarily the substitution of oil products as a means of energy storage.

All the evidence is that an increasing proportion of total energy consumption in the future will be supplied by electricity and the demand for both central energy storage and distributed local storage has to be met.

The diversification of the fuel base of transport is an important goal for industrial nations to pursue, and the prospective saving of oil warrants a major R & D effort on electric vehicles and traction batteries in parallel with research on synthetic fuels for internal combustion vehicles.

The world wide effort on energy storage research has increased rapidly during the last few years both in industry and in the public sector (see *Appendix C*).

Appendix A: Energy Tables

Work done by processes in nature can be described as transfer of quantities between two potentials

$$|\delta W| = |\Delta P \delta K|$$

or the creation or destruction of a quantity at a single potential

$$|\delta W| = |P \delta K|$$

A list of quantities and potentials from some basic processes is given in *Table A.1*.

Table A.1

Type of basic process	*Potential*	*Quantity*	*Conserved/ Not Conserved*	
Mechanical	Force (F)	Length (s)	C	
Kinetic	$v^2/2$ (v = velocity)	Mass (m)	C	
Gravitational	Gravitational potential ($g\,h$)	Mass (m)	C	
Spatial	Pressure ($-p$)	Volume (V)	C	
Elastic	Stress tensor	Strain tensor ('length')	C	
Thermal	Temperature (T)	Entropy (S)	C NC	Rev. processes Produced in irrev. processes
Chemical	Chemical potential (μ)	Number of moles (n)	C NC	Physical Chemical reactions
Electric	Electric potential (V)	Electric charge (q)	C	
Dielectric	Electric field (E)	Dielectric polarisation (P)	NC	
Magnetic	Magnetic field (H)	Magnetisation (M)	NC	

Table A.2 Storage capacity

	Energy densities (Wh/kg)	(Wh/dm^3)	Temperature (°C)
Conventional fuels			
Oil	11 000	8 300	
Coal	8 300	12 500	
Wood	4 200	2 000	
Heat storage (Thermal storage)			
Hot water	58	58	20–100
Hot rocks, (concrete, granite)	11	29	20–100
Iron ore	11	57	20–350
Aluminium	12	34	20–350
Iron	6	50	20–350
Heat storage: Phase change			
Ice, (heat of fusion)	93	93	0
Water, (evaporation)	630		100
Paraffin, (heat of fusion)	47	39	Approx. 55
Salt hydrates, (heat of fusion)	55	80	30–70 typical
Lithium hydride LiH, (heat of fusion)	1 300	1 070	686
Lithium fluoride LiF, (heat of fusion)	290	760	850
Fluorides, mixtures	200	400	450–850 conceivable
Lithium sulphate, Li_2SO_4	58	120	575
Ammonia Tiocyanat, NH_4SCN	12	16	88
Chemical storage			
Synthetic fuels:			
Ammonia	5 150	3 990	
Hydrogen (liquid)	33 000	2 490	
Hydrogen, metal hydride	600–2 500	2–5 000	
Ethanol (liquid)	7 694	6 100	
Methane (liquid)	1 389	5 900	
Methanol	5 800	4 400	
Propane	12 800	7 000	
Petrol	11 600	9 000	
Hydrazine	2 300		
Batteries:			
Lead–acid	40 (167)	80	20–30
Nickel–cadmium	100 (45)	100	
Iron–nickel	60 (266)	–	
Nickel–zinc	90 (321)	–	20–30
Sodium–sulphur	150 (680)	–	300–375
Lithium–sulphur	150 (1500)	–	
Iron–air	80 (70)	80	40
High temperature	400 (200)	400	350–450
Mechanical storage			
Water, pumped-hydro (100 m)	0.3	0.3	
Flywheels, steel	20–30	300–400	
Steel spring	0.1		
Natural rubber	8		
Electric fields			
Capacitor	–	2×10^{-4}	
Magnetic fields			
Superconducting coil	1–2	5–10	

Theoretical values in brackets. The storage is at standard temperature and pressure except where indicated.

Table A.3 Energy units

	J	*kcal*	*kWh*	*MWy*	*Btu*	*Mtoe*
1 J (joule) =	1	238.9×10^{-6}	278×10^{-9}	31.71×10^{-15}	947.8×10^{-6}	23.9×10^{-18}
1 kcal (kilocalorie) =	4186.8	1	1.163×10^{-3}	132.8×10^{-12}	3.968	10^{-13}
1 kWh (kilowatt hour) =	3.6×10^{6}	860	1	114.2×10^{-9}	3412	86×10^{-12}
1 MWy (megawatt year) =	31.54×10^{12}	7.53×10^{9}	8760×10^{3}	1	29.89×10^{9}	753.4×10^{-6}
1 Btu (British thermal unit) =	1055.1	252.1×10^{-3}	293.3×10^{-6}	33.46×10^{-12}	1	25.22×10^{-15}
1 Mtoe (Mill. tons oil equiv.) =	41.87×10^{15}	10^{13}	1.163×10^{10}	1328	3.968×10^{13}	1
1 ton pit coal contains	29.3×10^{9}	7×10^{6}	8150	930×10^{-6}	27.8×10^{6}	7×10^{-7}
1 m^3 natural gas contains	35.2×10^{6}	8400	9.8	1.12×10^{-6}	33.3×10^{3}	8.4×10^{-10}
1 ton crude oil contains	41.9×10^{9}	10^{7}	11650	1330×10^{-6}	39.7×10^{6}	10^{-6}
1 kg ^{235}U (total fision) contains	82×10^{12}	19.6×10^{9}	22.8×10^{6}	2.60	77.72×10^{9}	1.96×10^{-3}

The indications of the energy content of fossil fuels are average figures, as a variation exists according to the quality of the source. The variation is usually less than ± 5 %. At distillation, the energy content of oil rises in proportion to crude oil. Light oil rises 8 % and petrol 11 % on a weight basis.

Table A.4 Power units

	W	*hp*	*kcal/y*
1 W (watt) =	1	1.341×10^{-3}	7530
1 hp (horse power) =	745.7	1	5.615×10^{-6}
1 kcal/y (kilocalorie per year) =	132.8×10^{-6}	178.1×10^{-9}	1
1 ton pit coal/year =	930	1.25	7×10^{6}
1 m^3 natural gas/year =	1.12	1.50×10^{-3}	8400
1 ton crude oil/year =	1330	1.78	10^{7}
1 kg ^{235}U/year =	2.60×10^{6}	3486	19.6×10^{9}

Conversion to Mtoe/year, Btu/year etc. is by multiplying the value for kcal/year with the conversion factor from kcal to the unit in question (second line in Table A.3).

Table A.5 Prefixes

The following prefixes may be used to construct decimal multiples of units.

Multiple	*Prefix*	*Symbol*	*Multiple*	*Prefix*	*Symbol*
10^{-1}	deci	d	10	deca	da
10^{-2}	centi	c	10^{2}	hecto	h
10^{-3}	milli	m	10^{3}	kilo	k
10^{-6}	micro	μ	10^{6}	mega	M
10^{-9}	nano	n	10^{9}	giga	G
10^{-12}	pico	p	10^{12}	tera	T
10^{-15}	femto	F	10^{15}	peta	P
10^{-18}	atto	a	10^{18}	exa	E

Appendix B: Manufacturers of Energy Storage Equipment

Manufacturers, Electrochemical Storage and Conversion Devices (Batteries and Fuel-Cells)

Denmark

A/S Accumulator-Fabrikken,
Lyacvej,
2800 Lyngby,
Denmark

England

Chloride Group Ltd,
52 Grosvenor Gardens,
London SW1W 0AU,
England

Chloride Silent Power Ltd,
Davy Road,
Astmoor,
Runcorn, Cheshire WA7 1PZ,
England

Lucas Industries Ltd,
Great King Street,
Birmingham B19 2XF,
England

France

Citroën,
117 Quai André Citroën,
75747 Paris-Cedex 15,
France

Compagnie Générale d'Electricité,
54 rue la Boetie,
75382 Paris-Cedex 08,
France

Société des Accumulateurs Fixes
et de Traction (SAFT),
119 rue du Président Wilson,
92300 Levallois-Perret,
France

Italy

FIAT,
Corso Marconi 10,
Turin,
Italy

Magneti-Marelli Spa.,
Via Goastalla 2,
20122 Milano,
Italy

Japan

Japan Storage Battery Co. Ltd.,
Nishinosho,
Kisshoin,
Minami-ku,
Kyoto,
Japan

Matsushita Electrical Industrial Co.,
1006 Kadoma,
Kadoma,
Osaka-Pref,
Japan

Sony Corporation,
7–35 Kitashinagawa 6-Chome,
Shinagawa-ku,
Tokyo,
Japan

Toyota Motor Co Ltd,
1 Toyota-Cho,
471 Aichi,
Japan

Yuasa Battery Co.,
6–6 Josae-cho,
Takatsuke, Osaka-Pref 569,
P. O. Box 10,
Takatsuki,
Osaka
Japan

USA

Communications Satellite
Corporation, (COMSAT),
950 Lenfant Plaza,
SW Washington D.C. 20024,
USA

Dow Chemical Co.,
2020 Dow Center,
Midland,
Michigan 48640,
USA

Eagle-Picher Industries, Inc.,
P. O. Box 47,
Joplin, Missouri 64801,
USA

Electric Storage Batteries Inc.,
5 Penn Center Plaza,
Philadelphia, Pa. 19103,
USA

Energy Development Associates,
1100 W. Whitcomb Avenue,
Madison Heights, Michigan 48071,
USA

Energy Research Corp.,
3 Great Pasture Road,
Danbury, Connecticut 06810,
USA

Exxon Enterprises Inc.,
P. O. Box 45,
Linden, New Jersey 07036,
USA

Ford Motor Co.,
American Road,
Dearborn,
Michigan 48127,
USA

General Electric Co.,
1 River Road,
Schenectady, New York,
USA

General Motors Corp.,
GM Technical Center,
Warren, Michigan 48090,
USA

Gould Inc.,
30 Gould Center,
Rolling Meadows, Illinois 60008,
USA

Tyco Labs. Inc.,
16 Hickory Drive,
Waltham,
Massachusetts 02154,
USA

Westinghouse Electrical Co.,
Westinghouse Building,
Pittsburg,
Pennsylvania 15222,
USA

United Technologies Corp.,
410, Main Street,
East Hartford,
Connecticut 06118,
USA

Yardney Electric Corp.,
82, Mechanic Street,
Pawcatuck, Connecticut 02891,
USA

West Germany

Brown Boveri et Cie AG.,
62, Neustadter Strasse,
6800 Mannheim,
West Germany

Daimler–Benz AG.,
Postfach 202,
Mercedesstrasse,
7000 Stutgart-60,
West Germany

Deutsche Automobil GmbH
(DAUG),
Emil-Kessler-Strasse,
7300 Esslingen,
West Germany

Siemens AG.,
50 Werner-von-Siemens-Strasse,
8520 Erlangen,
West Germany

Fa. VARTA AG.,
Postfach 4280,
Dieckstrasse 42,
5800 Hagen,
West Germany

Manufacturers, Hydraulic-Pneumatic Accumulators

Christie Hydraulics Ltd,
Sandycroft Industrial Estate,
Chester Road,
Sandycroft, Deeside,
Clwyd, CH5 2QP,
Great Britain

Fawcett Engineering Ltd,
Dock Road South,
Bromborough,
Cheshire,
Great Britain

Greer Hydraulics, Inc.,
5830 West Jefferson Boulevard,
Los Angeles, California 90016,
USA (Canada, Israel, Mexico)

Olaer Industries S.A.,
z. Ind. 16 Rue de Seine,
92 Colombes, Paris,
France (Holland, Luxembourg)

Hydac GmbH,
Postfach 17,
6603 Sulzbach/Saar,
West Germany (Austria)

Nakamura Koki Co. Ltd,
1176 Nozato-cho Mishiyodo
gawa-Ku,
Osaka,
Japan

Olaer Italiana SPA,
10132-Torino,
Via Gassino 10.12,
Italy

Olaer Suisse,
Case Postale No. 305,
1701 Fribourg,
Switzerland

Olaer Espańa S.A.
Calle Moyanes,
16Y18, Barcelona 4,
Spain

Olaer Belge IRE. S.A.,
202 Rue Picard,
1020 Brussels,
Belgium

Appendix C: Research Programmes and Organisations

The major part of research and development (R & D) on energy storage is carried out within industrial companies throughout the world. In recent years, however, a considerable R & D effort has been established within the public sector. National programmes supported by governments in Europe, USA and Japan together with international programmes initiated by OECD in Paris and by EEC in Brussels have emerged.

International cooperation on energy conservation and solar energy includes research and demonstration programmes on energy storage. The two largest international cooperational bodies are:

1. International Energy Agency, OECD, with participation from all industrialised nations except France
2. Commission of the European Communities – Energy Research and Development Programme – section h: development of methods for storage of secondary energy, with participation of several EEC member states.

In the USA the activities of the former Energy Research and Development Administration (ERDA) is now part of the new governmental Department of Energy (DOE). R & D is carried out at universities, national laboratories and as contract work with private companies. The energy conservation programme of the DOE has a special Division of Energy Storage Systems.

In Japan public support of energy storage R & D is administrated by the Agency of Industrial Science & Technology under the Ministry of International Trade & Industry (MITI).

There is a tendency to focus most attention on industrial projects rather than fundamental research. This is true both for the above mentioned organisations and for most of the European governmental programmes set up on a national basis.

The objective of the public research programmes on energy storage especially in the USA and in Japan has in the period from the late 1960s until 1974 been to reduce air pollution. More recently the major concern has been in the field of energy conservation and in particular the reduction of dependence on oil by introducing non oil based energy storage systems.

Index